PLACEBO

OTHER MERIDIAN TITLES

Howard Bridgman, Robin Warner, and John Dodson, *Urban Biophysical Environments*

David Chapman, *Natural Hazards*, Second Edition

Arthur and Jeanette Conacher, *Rural Land Degradation in Australia*

Robert H. Fagan and Michael Webber, *Global Restructuring: The Australian Experience*, Second Edition

Dean Forbes, *Asian Metropolis: Urbanisation and the South-East Asian City*

Nick Harvey, *Environmental Impact Assessment: Procedures, Practice, and Prospects in Australia*

Iain Hay, *Communicating in Geography and the Environmental Sciences*

Clive Forster, *Australian Cities: Continuity and Change*, Second Edition

Jamie Kirkpatrick, *A Continent Transformed: Human Impact on the Natural Vegetation of Australia*, Second Edition

Elaine Stratford, *Australian Cultural Geographies*

PLACEBOUND

AUSTRALIAN FEMINIST GEOGRAPHIES

LOUISE JOHNSON

WITH

JACKIE HUGGINS AND JANE JACOBS

SERIES EDITORS
DEIRDRE DRAGOVICH
ALARIC MAUDE

OXFORD

UNIVERSITY PRESS

OXFORD

UNIVERSITY PRESS

253 Normanby Road, South Melbourne, Victoria, Australia 3205

Oxford University Press is a department of the University of Oxford.
It furthers the University's objective of excellence in research.
scholarship and education by publishing worldwide in

Oxford New York
Athens Auckland Bangkok Bogotá
Buenos Aires Calcutta Cape Town Chennai
Dar es Salaam Delhi Florence Hong Kong
Istanbul Karachi Kuala Lumpur Madrid
Melbourne Mexico City Mumbai Nairobi Paris
Port Moresby São Paolo Singapore Taipei
Tokyo Toronto Warsaw

and associated companies in Berlin Ibadan

OXFORD is a trade mark of Oxford University Press
in the UK and certain other countries

National Library of Australia
Cataloguing-in-Publication data:

Johnson, Louise, 1953– .
  Placebound: Australian feminist geographies.
  Bibliography.
  Includes index.
  ISBN 0 19 553566 9.

1. Feminist geography – Australia. 2. Feminist theory –
Australia. 3. Women's studies – Australia. 4. Geography –
Philosophy. 5. Human geography – Australia. I. Title.
(Series: Meridian, Australian geographical perspectives).

305.4201

Edited by Gillian Fulcher
Indexed by Russell Brooks
Text designed by Polar Design Pty Ltd
Cover designed by Polar Design Pty Ltd
Typeset by Polar Design Pty Ltd
Printed through Bookpac Production Services, Singapore

Foreword

Meridian: Australian Geographical Perspectives is a series initiated by the Institute of Australian Geographers in 1990 to meet the need for relatively short, low-cost texts written for university students. The books in this series are designed to explore the geographical issues and problems of Australia and its region, or to present an Australian perspective on global geographical processes. The term meridian refers to a line of longitude linking points in a half-circle between the poles. In this series it symbolises the interconnections between places in the global environment and global economy, which is one of the key themes of contemporary geography. The books in the series cover a variety of physical, environmental, economic and social geography topics, and are written for use in first and second year courses where the existing texts and reference books lack a significant Australian perspective. To cope with the very varied content of geography courses taught in Australian universities the books are not designed as comprehensive texts, but as modules on specific themes which can be used in a variety of courses. They are intended for either a one-semester course, or a one-semester component of a full year course.

Titles in the series cover a range of topics representing contemporary Australian teaching and research in geography, such as economic restructuring, vegetation change, land degradation, cities, cultural geography, natural hazards, and urban environmental problems. Future topics include qualitative methods and coastal management. Although the emphasis in the series is on Australia, we will also have occasional titles on Southeast Asia, using the considerable expertise that Australian geographers have developed on this region.

While the primary aim of the series is to produce books for students, the topics selected deal with issues of relevance to all Australians. We therefore hope that the general reader will find some of the titles of interest, and discover that geographers have something distinctive to say on contemporary environmental, economic and social issues. As the books assume little or no previous training in geography, and attempt to avoid a textbook style, they should be readily understood by the general reader.

Louise Johnson's book, with a contribution from Jackie Huggins and Jane Jacobs, is the eleventh in the series. It introduces the reader to the field of feminist geography, which highlights the neglect of women in and by a male-dominated discipline, and shows how gender influences geography in the ways in which women use, react to, and are influenced by places and space. The book examines the often subordinate position of women within these places, and the ways proposed by different groups of women to alleviate this problem. The book situates feminist geography in the various strands of feminist thinking, from liberal feminism to socialist and radical feminism, and then considers the impact on feminist geography of postmodernism and postcolonialism. This enables the reader to see the links between feminist theory, contemporary social theory, and feminist geography, and the ways in which the category of woman becomes more and more complex as class, ethnicity, race and sexuality are incorporated into the discussion. The book should therefore be of interest and value to a wide range of readers interested in feminism, and not only to geographers.

Louise Johnson has been researching and teaching feminist geography for over a decade, and she brings much of her personal experience to this book. The reader will learn much from her applications of feminist geography to such topics as mapping the status of women, local restructuring in a patriarchal economy, 'cities of fear', sexuality in the suburbs, and the postmodern shopping centre.

Deirdre Dragovich
University of Sydney

Alaric Maude
Flinders University

Contents

List of Figures

List of Tables

Acknowledgments

The intellectual origins of this book lie with the Geography Department at New Zealand's Waikato University. Here, in 1987 I was invited to teach an undergraduate course in Feminist Geography at a time when any such notion was unheard of in Australia. My colleagues, friends and the students at Waikato, in particular Terry Baxter, Louise Dooley, Neil Ericksen, Beryl Fletcher, Richard Le Heron, Noelene Ryan, Evelyn Stokes, Susan Sayer and Anne Sullivan, all stimulated and refined many of the ideas which now form the content of this book.

But turning a course into a book also takes an invitation—and for this I thank Stephanie Fahey and Alaric Maude—and a great deal of time, thinking and work. For the space in which to develop the ideas in the book I owe a special thanks to the faith, encouragement and patience of Jill Henry at Oxford University Press and Joan Beaumont at Deakin University. For the challenges, active assistance and support which have been necessary for its completion I thank Donna Ferretti, Rodney Harris, Lesley Instone, Kate Kerkin, Vivienne Milligan and Lesley White. But for the domestic sustenance, love and endless understanding which a five-year project involves I owe most to Sue, Jarred, Natham, Sian and Liam.

Abbreviations

ABS	Australian Bureau of Statistics
CBD	Central Business District
CIA	Central Intelligence Agency
CR groups	consciousness-raising (groups)
DJs	David Jones
EEO	Equal Employment Opportunity
FBI	Federal Bureau of Investigation
GRC	Geelong Regional Commission
HALCS	Housing and Locational Choice Survey
IHC	International Harvester Company
RGS	Royal Geographical Society
RGSA Q	Royal Geographical Society of Australia, Queensland branch
UAW	Union of Australian Women
US	United States (adjective)
USA	United States of America

Introduction

Are there spaces in Australia that are quintessentially men's spaces? Or women's spaces? Or gay spaces?

This question, asked some years ago of a group of undergraduate students, initially brought consternation and a long silence: in 1988 it was a question rarely asked. But after the silence, a list emerged, and from it the conclusion that women and men inhabit quite different environments even if they live side by side. Places are gendered, that is, they acquire and evince certain characters because they are occupied by men or women. The football field, the corner pub, an engineering shop or a board room are male spaces: they differ from their female counterparts: spaces like the netball court, a child-care centre, checkouts at a supermarket or the home interior. A darkened street takes on a gendered quality as soon as it is approached by a woman or a man. Thus in their design and use places declare their masculinity or femininity, and their value and power. Women's spaces tend to have low status and to be associated with limited political and economic potency. Even though men's spaces are not uniformly those of wealth and influence, men still tend to monopolise powerful places. Thus spaces register gender, and reinforce patriarchy, a gender order which privileges and advantages men over women[1].

This simple dichotomy contains great complexity. Between men who labour in an engineering shop and men who sit in the manager's office there is a huge class divide. The professional woman and the single working-class mother who use child-care centres have different class and household situations; they may also have different ethnic and racial backgrounds, may speak languages other than English, and may live with male or female partners. Thus class, ethnicity, household form, race, language and sexuality variously define women and men. They also impinge on the spaces they inhabit and affect the character of those places. Some geographers—'feminist geographers'—are focusing on these gendered spaces, with radical implications for the nature of the discipline and the environments they study.

Geography is the study of the relations between people and their environment. Over its history the discipline has created various traditions of enquiry: these include earth science, spatial relations, the connections between humans and their environments, and the systematic examination of regions or places (Pattison 1964). In each of these, the primary focus has been on worlds now seen as masculine: the physical environment, the economy and public spaces like cities and factories; and the approach to their study was mainly scientific. As female geographers became aware of this bias, they also recognised that it was mainly men who created and oversaw the discipline. Thus in the early 1980s, it came to be seen that geography had sexist biases in its content, methods and purpose (Monk and Hanson 1982, p. 14). This realisation, and the critique and reconceptualisation which followed, has produced a revolution in geographical thought, pedagogy, methods and departments. The nature of this revolution is the subject of this book.

Inspired by a feminist politics, geographical space has come to be seen not as a neutral container of social and bio-physical relations but as a medium which registers and expresses power and sexual difference. The study of these gendered spaces, as well as moves to alleviate the subordinate position of women within them, is the concern of feminist geographers. First emerging in English-speaking countries in the early 1970s, Australian feminist geography has drawn perspectives and techniques primarily from British and North American geography as well as from feminist scholarship more generally. This work, and my knowledge of it, derives primarily from work produced in the Western and English-speaking world, where enquiry has now broadened to include research both on and by women from non-English-speaking countries, as well as a postcolonial critique of the discipline.[2] In the interests of manageability and within the confines of my expertise, the focus of this book will primarily be on the feminist geography that has emerged in English-speaking Western countries.

In feminist geography, the initial focus on the scarcity and lack of power of female geographers has persisted along with political activity to increase the presence and visibility of women in the discipline. There has also been an ongoing critique of geographical writing and practice, as well as moves to reveal the place of women in the landscape. This liberal feminist orientation has been joined by other concerns. Thus, for those trained in Marxism, their feminist consciousness has often led to socialist feminism, and to a focus on women in the paid and unpaid workforce in particular locations. More recently, some feminist geographers have returned to an earlier tradition in women's studies, namely radical feminism, where they focus on the regulatory power of patriarchy as well as sexuality in the making and meaning of spaces. Contemporary developments in social theory, which are loosely encompassed by the labels postmodernism and postcolonialism, have directed attention to new sites of interest, especially the body, language and the politics of racial, sexual and ethnic difference.

The book's aim is to show how each of these theoretical territories has shaped feminist geography, and how they have been used in studies of particular Australian environments. The book thus contains a number of theoretically informed studies of spaces which are not only characterised by gender but also by class, race, and sexuality. I shall also use these case studies as vehicles for critically engaging with various theoretical, methodological and empirical positions in feminist geography.

BOOK OUTLINE

Emanating from the Women's Liberation Movement of the 1960s and 1970s, gender awareness in geography has primarily meant that female geographers have recognised their marginalisation in the discipline and, more generally, women's marginalisation in constructed environments. This has led to work which reveals the many ways in which men express their power in particular spaces. Partly in response to feminist questioning, some male geographers began to look critically at themselves and their role in changing the face of the earth. But few male geographers have considered 'men's spaces'[3], even though they are fundamental to geography and to the organisation and regulation of Australian environments. The book's main concern, however, is not with men's spaces but with women—those most usually excluded from geographical knowledge—their spaces and the range of perspectives for the study of those spaces.

Chapter 1 introduces liberal feminism and then critically examines 'liberal feminist geography'. The focus on individual rights and gender equality, and the possibility of positive change through the removal of institutional barriers, has led geographers in a number of directions. Some of this work has revealed the number of women in the discipline—as students and researchers, as senior members of academic, administrative and publishing communities—and has been part of attempts to establish equal opportunity and affirmative action policies. Other research has described the location of gendered groups and has begun to map the different locations of women and men. As well as documenting women in Australian geography, the chapter uses a status of women index (first applied by North American geographers in the 1970s) to illustrate women's unequal position with men in Melbourne and Victoria. It also outlines more recent work on social polarisation in the city, which has built on concerns for social justice and equity, enriched the earlier focus on women as an undifferentiated group, and given the liberal feminist perspective new directions for empirical study. The chapter points to the political and conceptual strengths as well as the limitations of the framework.

Some female geographers found the liberal feminist framework lacking, particularly those who, first trained in Marxist theory, found liberalism inadequate to explain unequal environments. While Marxism is a diverse tradition (see Johnson 1990a), socialist feminist geographers have tended to explore the two key categories in socialism—those of production and social reproduction—in order to understand women's paid and unpaid work in urban and regional environments. Chapter 2 provides an overview of Marxism and socialist feminism, and a critical survey of 'socialist feminist geography'; it also describes a study of one region which has undergone major restructuring over the last twenty years, namely Geelong in Victoria. Here I draw on my own version of socialist feminist theory, arguing for the centrality of the patriarchal economy to understanding this locality as Geelong's manufacturing industries contract and parts of its service sector expand. Thus the new sexual divisions of labour in these workplaces and, relatedly, in the home, may be described as new relations of production and reproduction.

In this kind of regional and workplace study, the importance of global, national and local economic relations is critical. For socialist feminists, it is these dynamics which ultimately structure the patriarchal economy. Thus much socialist feminist geography assigns more importance to Marxist than to feminist categories. It is partly in reaction to such a ranking as well as the articulation within geography of a feminist tradition long present in women's studies—that of radical feminism—which put patriarchy at the centre of analysis. This is especially apparent in research on 'cities of fear', and in work which has taken sexuality as the key dimension in a person's gendered identity. Thus radical feminist geography places male power, gender and sexuality at its centre and concentrates on the power dynamics and spaces which emerge from these dimensions. While most geography on sexuality has focused on marginal groups, such as gay men and lesbians, a consideration of the Australian city in terms of heterosexuality elucidates both the ways in which this category is defined and maintained in suburban areas, and how it is challenged by gay geographies and queer theory. In exploring sexuality in the city, I shall use case studies from an Adelaide suburban development, and Sydney's Gay and Lesbian Mardi Gras. This section looks at inner city spaces and the suburbs and domi-nant, marginal and disruptive sexualities. The strengths and the limitations of such an approach are also outlined in chapter 3.

Many who have put sexuality at the centre of their geographies have noted the connection between identity formation and more general cultural and economic developments. The new cultural geography has emphasised that this connection is related to a major change which allegedly began in the 1970s in the global and local economies. Theorists such as Harvey (1989a), Jameson (1984) and Soja (1989) have seen the shift from a Fordist form of organised capitalism to a post-Fordist or disorganised regime of flexible accumulation as a transition to post-modernism. This view of cultural and economic change brings a scepticism towards all-encompassing explanations, a greater emphasis on language and discourse, a concern for the body, and, in 'postmodern geographies', an emphasis on consumption, the spectacle and identity. In feminist geography, this view has led to a heightened interest in culture, representation, consumption and the gendered body in space. The regional shopping centre is often cited as a place where the economic and the cultural merge in postmodern ways, and where the politics of spectacle, consumption and embodiment are played out. In chapter 4, one such centre in Sydney, namely Westfield Shoppingtown in Parramatta, will provide the vehicle for exploring 'postmodern feminist geography'.

The postmodern geographer's focus on contemporary consumption palaces aims to highlight the nature of the economy supporting those activities and to elaborate the identities being constituted in these places. This exercise can also identify the ways in which global capital and culture intersect locally. This concern with the local, the particular and the vernacular is one dimension of the postmodern, but it has also informed another position in contemporary geography, namely postcolonialism. Thus imperial history, and locating the discipline and places in that history, is one element in contemporary feminist geography. Here, the emphasis is on the nature of difference between localities and those who live in them. Such discussions focus on positionality, that is, where each person is located, how they speak, who they speak for and from what location within a skein of past and ongoing imperial power relations. Thus 'postcolonial geography' seeks to uncover the ways in which colonisation continues through the elevation of white, Western experience, and through the objectification and homogenisation of 'others', whether places or peoples. The intention in postcolonial geography is thus to undermine the authority of the

West to speak for others, and to give a voice to those long silenced by its imperial dominance. In particular, postcolonial geography is concerned with how and who feminist geography has silenced or made marginal.

Liberal feminism assumed that women were relatively undifferentiated, united in their oppression by men and adequately represented by those white middle-class women who were the voice of liberal reform. This view of sisterhood carried over into radical feminism, though a recognition of women's different sexualities identified a major differentiating element between women. For socialist feminists, however, women were constituted by class relations as much as by gender, though ethnicity and race were also admitted as complicating elements in defining a class position. More recently, other voices have demanded that established feminists in Australia, North America and Britain recognise that their work has primarily advanced the interests of white Anglo-Celtic and middle class women. In Australia, these demands are further complicated by the question of indigenous rights. Few of these voices have been heard directly, and it was for this reason that I approached Jackie Huggins to write a section on Aboriginal spaces for this book.

The fraught nature of this request was soon revealed. Chapter 5 pivots on this finding. I asked both Jackie Huggins and Jane Jacobs, whom she chose as her co-writer, to write as self-consciously Aboriginal and white authors. Their brief was to make the issue of difference from an articulated norm of a white, Anglo-Celtic woman central to their writing; at the same time, they were to offer a case study from the point of view of their race and ethnicity. I had assumed that race and gender created a unitary and unidimensional identity. As their account makes clear, however, no such assumption can be made. Thus gender, like race, class and ethnicity is a category which is not fixed but is the subject of negotiation, imposition, contestation and change. Furthermore, how accounts about the colonial experience or about a particular place are written involves recognising the differences which make a spatial history of this kind. What thereby emerges from this chapter is a sense that there are indeed differences between women as there are between women and men and that such differences are variously derived from power imbalances which are inscribed onto space. Who defines such positions and shapes the imbalances is a complex and historically contingent question without easy resolution.

Chapter 5 brings to the fore the issue of positionality and the construction, regulation and consumption of places. It does so using the example of Carnarvon Gorge in Central Queensland which is examined from two particular points of view; these viewpoints are both defined by race, class, gender, ethnicity, history and education and go beyond these dimensions. Kooramindanjie Place or

Carnarvon Gorge is constituted by the various racial and sexual meanings which are attached to it, and the power dynamics which cohere around these dimensions. Here geography is as much about race and colonialism as gender; it is also about the politics of representation as well as about place. The subject and the approach therefore raise significant questions for feminist geographers in Australia and elsewhere about how they write and from what position. It also raises a fundamental question: has a feminist geography concentrating solely upon gender relations in space a future? Postcolonial thinking shows the impossibility of a notion of a unified group 'women': there can only be black or English speaking or Islamic or...women, or whatever qualifiers are appropriate. Gender identity, then, like spaces and places dissolves into a fluid and contested array of discourses, directions and political interventions. This, then, is the uncertain and challenging future for Australian feminist geography, and the topic of chapter 6.

In this book, the various traditions which constitute feminist geography in Australia are critically presented and applied to particular case studies. These traditions show that Australian environments are both gendered and constituted by relations of class, sexuality and race. They also show that, in various ways, women are Placebound. In many ways, women are bound to particular places, such as the home, low level jobs in the paid workforce or trudging through the aisles of shopping centres; as they are to an array of social places: as heterosexuals, 'good' mothers and workers, as Black women, as sexualised bodies. Despite these confinements women are also breaching these boundaries, making new places of their own, whether in the discipline of geography, in the public or the private spheres. So women are both bound to place but also moving towards their own places—Placebound—bound by their own energy, critiques, differences and creativity.

Notes

1   The term patriarchy is one which has been the subject of much debate in feminist literature and in feminist geography. Its meaning is therefore contested and is only tentatively fixed here. The 'patriarchy debate' is examined in chapter 3.

2   Chapter 5 looks at the impact of the postcolonial critique on feminist geography and further explores the idea of positionality, that is, the question of where an author writes from, sociologically and ideologically. I have approached this book as a white, monolingual, Anglo-Celtic, urban, middle-class feminist geographer; the limitations of my knowledge derive from my stated identity markers (though many other dimensions could have been mentioned) and from my training in a metropolitan Australian university system which has only very recently started to recognise the existence of countries and literatures beyond the Western, English-speaking universe. In feminist geography, these

limitations are pervasive, though the International Geographical Union's 'Geography and Gender Study Group' and the more European orientation of British feminist geographers have begun the process of de-Westernising and broadening feminist geography. In this book I have located Australian scholarship and places in various global networks; chapter 5, especially, recognises the interconnection between First- and Third-World economies and societies. Regretfully, the substantial feminist literature in the area of women in Third-World countries (see, for example, Momsen and Townsend 1987; Thomas and Skeat 1990) is outside my expertise and will not be considered here.

3   There are very few male academic geographers who actively recognise and attempt to interrogate their own and the position of other men in space. Some of the more notable exceptions are Peter Jackson (1991), and in gay geography the work of Manuel Castells (1983), Glen Elder (1995), Lawrence Knopp (1990, 1992) and Micky Lauria (Lauria and Knopp 1985). Gay geography is considered further in chapter 3.

1

Liberal Feminist Geography

The Women's Liberation Movement of the 1960s and 70s was the necessary precursor of feminist geography. This movement emerged from a range of demographic and social changes. These changes included the maturing of postwar baby boomers, the mass entry of married women into paid work and higher education, political movements demanding youth, gay and minority rights and a revolution in contraceptive technology which freed sexual activity from procreation (Mitchell 1971).

A profoundly conservative postwar decade preceded these developments. Writing of America in the 1950s, Betty Friedan described the anxiety which many white, middle-class women felt about their affluent but stultifying lives, as the 'problem with no name'. Their lives were intensely suburban, revolving around home making, child rearing and heterosexuality (Friedan 1963). This was the dominant image and reality for these women and it contrasted with the wider society in which there were increasing opportunities in education and paid work, and a political landscape which included student activism, anti-Vietnam War protest, calls for Black power, and demands emanating from the 'New Left', for radical social change (Mitchell 1971).

In Australia, there were similar contradictory processes at work. After 1945, women who had been involved in war work were encouraged (as well as forced) to leave their well-paid jobs, return to their homes, heal their battle-weary men and rear the next generation. Post-war scarcities made home making a trial, while memories of war-time autonomy, paid work and economic equality contrasted with admonishments from 'experts' that women be perfect mothers, home makers and consumers (Matthews 1984). It was in this climate that, in 1950, the New Housewives' Association moved beyond a concern with domestic issues to form the Union of Australian Women (UAW). Initially, the UAW focused on world peace, human rights and rising prices but as the decade progressed the organisation focused on women's issues: it campaigned for maternity allowances, child care, increases in child endowment and equal pay. Through such actions, women became aware of the connections between their private domestic concerns and the public sphere.

The fight for equal pay became the catalyst for the beginning of Women's Liberation in Australia. Women took collective action, seeking to overturn the 1907 decision in the newly created Commonwealth Arbitration Court which was known as the Harvester Judgment. In this judgment, Justice Higgins presumed that men were breadwinners supporting families and that women in the paid workforce needed an income only to maintain themselves. He concluded that women were entitled to only 54% of the basic male wage. This decision had enshrined differential pay rates for men and women; it was supported over the next 50 years by both employers and male trade unionists (Matthews 1984). But in the 1960s, in the climate of expanding employment opportunities and widespread political agitation, women organised to challenge this situation.

Thus in 1969, after a series of demonstrations by women demanding equal pay, a group chained themselves to the Commonwealth Building in Melbourne. The Arbitration Court subsequently granted equal pay for equal work; the male-dominated union response was passive, and these events stimulated the creation of more radical women's organisations. These new organisations urged women to take public action on their own behalf, and form women's groups which both advanced their interests and critically evaluated the subordinate role assigned to them in trade unions, Leftist organisations and other mixed groups. These radical groups also defined women's position not as inequality but as one of oppression (Gibbs 1985).

As well as these groups campaigning for equal pay, the Women's Liberation Movement of the 1960s included women who, angry and frustrated, actively protested and agitated for change; these women were impatient of carefully articulated manifestos and formal organisations. In this mobilisation, women created a political movement and new ways of conducting politics both in Australia and the USA. In the USA, 'Zap actions' targeted the Miss America pageant and sexist graffiti, and consciousness-raising (or CR) groups spread. Such groups borrowed from the Chinese Communist Party's technique of 'speaking bitterness', and small groups of women met in a supportive, egalitarian and confidential setting, to talk of their personal experiences. In the process of relating their experiences in such groups, personal stories moved from being idiosyncratic to shared. CR groups became the site where women articulated not what they were supposed to feel but what they did feel, a process of validation. These individual stories revealed the shared, systematic and political pattern of women's oppression (Eisenstein 1984). The 'problem with no name', which Betty Friedan had identified in 1963, began to be named and politicised. As Juliet Mitchell wrote:

> Women come into the movement from an unspecified frustration of their own private lives, find that what they thought was an individual dilemma is a social predicament and hence a political problem. The process of transforming the hidden, individual fears of women into a shared awareness of the meaning of them as social problems, the release of anger, anxiety, the struggle of proclaiming the painful and transforming it into the political— this process is consciousness-raising (Mitchell, 1971, p. 61).

In this exercise of identifying wider social structures of power and powerlessness, men were seen as the problem. Hence, women's collectives and coalitions organised political strategies and actions on issues such as men's violence, rape, abortion and equal pay. The ideology informing this activity became known as 'radical feminism' (see chapter 3). Radical feminism inevitably built upon its predecessor, that of liberal feminism.

Liberal feminism has existed for several hundred years; it remains central to contemporary feminism and it provided a critical theoretical framework for some of the earliest statements in feminist geography.

Thus, in a 1973 issue of *The Professional Geographer*, Wilbur Zelinsky wrote on 'The strange case of the missing female geographer' in the USA (Zelinsky 1973a). Noting the widespread attention to the Women's Liberation Movement and the existence of women's caucuses in a number of academic disciplines, Zelinsky portrayed geography as lagging behind; moreover, he pointed out that the few women in the discipline did not lobby in their own interests, either as individual women or for women in the discipline (Zelinsky 1973b). He saw this as a problem for these women, and both the discipline and the wider society were 'being cheated out of major quantities of intellectual achievement and skilled womanpower' (Zelinsky 1973a, p. 101).

Zelinksy's two articles were among the first in English on women in geography, and they owed much to liberal feminism in their formulation of, and proposed solution to, the problem. In Zelinsky's view, the problem for geography and the women in it was one of representation and visibility. Within a liberal political framework, the solution then became that of attitudinal, legal and institutional changes to encourage more women into the discipline, to map their presence in the landscape and to create an inclusive geography. These considerations have informed much of feminist geography in North America, the United Kingdom and Australasia.

While there is a long history of those writing about and analysing women's oppression in Western societies (Spender 1982), those who first did so, wrote from within a liberal viewpoint; they included Mary Wollstonecraft, John Stuart Mill and Harriet Taylor. Their liberal feminism has a number of conceptual and political strengths which have secured, and continue to secure, some important victories within geography and elsewhere. For instance, it directs attention to inequality and social justice; hence it remains a vital framework. But liberal feminism also has some profound weaknesses which limit the ways in which problems can be defined and solutions formulated. However, the position is still seen as a vital one—directing much needed attention to inequality and social justice— and it has been subject to critical reformulation by contemporary feminists such as Carole Pateman and Nancy Fraser. This chapter will present an overview of liberal feminism before sketching some of the ways in which it has directed contemporary geographers to question the number and status of women in the discipline, to formulate corrective policies and practices, and to create a new geography where gender inequality is mapped and women are made visible.

LIBERAL FEMINISM

Liberalism was a set of ideas which developed alongside the economic triumph of the European bourgeoisie in the seventeenth and eighteenth centuries. It comprised views which justified and sustained the wresting of political power from feudal overlords and its centralisation—along with the means of production—into the hands of a new middle class. It was also an ideology which inspired the ending of bonded labour and the creation of a working class whose members saw themselves as freely selling their labour for a just reward. Liberal principles also informed the creation of a series of racist Empires to provide necessary raw materials, foodstuffs and capital outlets to industrialising Europe. In addition, and significantly for later feminists, liberalism justified a gendered separation of home and the private sphere from the public world of paid work and politics. These notions derived from and belonged to the public realm of white men. The creation of liberal notions such as 'rights', 'choice', 'the individual', 'freedom', 'reason', 'justice' and 'equality' therefore derived from particular class, racial and gendered interests. Despite this origin, liberal principles have informed analyses of women's position, but because of this deviation they have also limited that analysis and the politics which flow from it.

Liberal feminism was first formulated in Mary Wollstonecraft's *The Rights of Woman* (1792; 1929) and in John Stuart Mill's (and Harriet Taylor's[1]) *The*

THE RIGHTS
of WOMAN *by*
Mary Wollstonecraft
& The SUBJECTION
of WOMEN *by*
John Stuart Mill

LONDON & TORONTO
PUBLISHED BY J·M·DENT
& SONS L.TP & IN NEW YORK
BY E·P·DUTTON & CO

Figure 1.1   Frontispiece to the 1929 edition of *The Rights of Woman* and
The Subjection of Women

Subjection of Women (1869; 1983). Wollstonecraft's work attempted to explain and
arrest the decline in women's status which accompanied the shift of productive
activity from the home into paid workplaces. This change firstly affected middle-
class women who, for Wollstonecraft, became emotional cripples, petty shrews and
narcissistic sex objects. As a result, women's powers of reason were stifled and they
were reduced to a privileged, enervated and emotional state. To counter this,
Wollstonecraft argued that women be more highly educated. This would develop
girls' powers of reason and their moral capacities to achieve personhood. While she
recognised that women should be economically and legally independent of men,
she was unclear how these things could be achieved. Wollstonecraft also saw no
contradiction between the systematic development and exercise of reason and the
discharge of wifely and motherly duties (Tong 1989).

Writing 100 years later, the influential British philosopher John Stuart Mill, and his collaborator Harriet Taylor, broadened the political agenda from education to one of women gaining a range of other civil rights. In *The Subjection of Women* (1869; 1983), Mill and Taylor exposed the irrationality of arguments for female subjection before locating them in the physical weakness associated with child bearing. They went on to question the conversion of these 'physical facts' into legal rights, identifying marriage as the key institution which transforms the physical frailty of women into dependence on men. Therefore, even though women may 'choose' to marry, this is not a free choice when there are no real alternatives. Further, within marriage, Mill and Taylor argue, there is an arbitrary inequity based on male power. Such power can only be countered when women can vote and have access to higher education and the professions. The consequences of such reforms would be women's happiness, for they would no longer have to deny themselves and could use their many talents for everyone's benefit. Men too would gain, for they could be less self-sacrificing and less selfish in families which were more just and free. Society would also be richer because of the extension of the liberal principles of equality and justice to all. The 'moral regeneration of mankind' would follow the removal of restrictions on women and the replacement of subjection by 'rational freedom' (Mill and Taylor 1983).

Mill and Taylor's argument was rational, egalitarian and liberal, and it informed popular campaigns for women's suffrage. Eventually, their argument and political agitation led to women becoming members of parliaments, having access to higher education and the professions, and to their securing inheritance and property rights. However, the liberal feminism which inspired such action assumed the continued existence of the bourgeois family and its traditional division of labour. There was also an unstated assumption that the white masculine subject was the ideal citizen, the model against which others were to be judged. Thus Mill and Taylor wrote: 'When the support of the family depends, not on property, but on earnings, the common arrangement, by which the man earns the income and the wife superintends the domestic expenditure, seems to me in general the most suitable division of labour between the two persons' (Mill and Taylor 1983, pp. 87–8).

There are a number of consequences which flow from Mill and Taylor's analysis and reforms. While they argue that a married woman should have rights to property, they limit these rights to property she herself has inherited or earned, not rights to equal shares in the family income. Further, inequality within the family, they suggest, arises from this differential in property, not from a larger sexual division of labour built upon women's lesser position in the paid labour force. In addition, even with their reforms in place, Mill and Taylor assumed that women would still choose to marry. Their analysis thus constructs a conservative view of women's aspirations and potential (Okin 1979;

Eisenstein 1981). What Mill and Taylor were advocating therefore was limited: tinkering with the effects rather than the fundamental causes of women's oppression. It was this view that socialist feminists challenged (see chapter 2).

In their critique of Enlightenment thinking, feminists such as Carole Pateman (1988), Genevieve Lloyd (1984), Judith Butler (1990, 1992) and Elizabeth Grosz (1988) have revealed further limitations in the liberal view of women. They argue that the subject which underpins liberal feminism is inherently masculine and racist, so that the female is always lesser and other to the white, male norm. Liberal feminism is thus unable to include a notion of woman as an autonomous, differentiated subject. Hence, the liberal feminist project is doomed to philosophical and political failure.

In *The Sexual Contract*, Carole Pateman argues that the liberal notion of free individuals coming together to create society affirms and universalises the attributes of men. She suggests that, as the archaic order which gave authority to fathers and patriarchs became obsolete, it was replaced by an implied contract between male citizens. This contract separated the private from the public sphere and established the subordination of women to men as the foundation of modern society. It also excludes women's concerns from moral and political consideration and banishes them to the realm of nature. Hence liberalism's patriarchal social contract and civic order cannot provide a basis for women to take an equal place in society (Pateman 1988; Curthoys 1994).

Writers such as Moira Gatens (1992) have further explored the idea of the liberal individual as masculine. Gatens sees Mill as constructing the masculine experience as the norm. Similarly, Carol Pateman sees the individual as explicitly associated with the attributes and roles of men as opposed to women: 'The masculine, public world, the universal world of individualism, rights, contract, freedom, equality, impartial law and citizenship...gains its meaning and significance only in contrast with and in opposition to, the private world of particularity, natural subjection, inequality, emotion, love, partiality—and women and femininity' (Pateman 1986, p. 6).

In a social order where women are identified with their bodies and emotions and men with reason and logic, this suggests that men are natural rulers and politicians. Rationality is thereby linked to masculinity in Western thought, or as Genevieve Lloyd writes: 'The obstacles to female cultivation of Reason spring to a large extent from the fact that our ideals of Reason have historically incorporated an exclusion of the feminine, and that femininity itself has been partly constituted through such processes of exclusion' (1984, p. x).

The Western philosophical and political tradition has therefore associated maleness with rationality, reason, individuality and citizenship, and with a disembodied rationality which uniquely qualifies men for public life (Bulter 1990,

1993). Men have been constructed as above and beyond the flux of emotion and bodily needs. In contrast, women have been cast as ruled by their bodies, by nature and by emotion. Against this view, Judith Butler argues that the 'natural-ness' of women's bodies is a contingent fact to be deconstructed, an exercise which reveals the unequal politics of gendering and of heterosexuality (Butler 1990, 1993. See chapter 3). Liberal philosophy constitutes the rational and disembodied masculine subject as the citizen and thus the subject for liberal geographers. Geographer Gillian Rose notes that, against this representation, not only is male embodiment an unacknowledged foundation for malestream thinking, but it is also a white heterosexual body which forms the foundation for liberal feminist geography, especially time geography (Rose 1993).

There are further fallacies in the liberal conception of gender inequality: its unitary notion of women as defined by their bodies, biology, irrationality and location in the private sphere, denies important social differences between women. This conception derives from those white, middle-class heterosexual men and women in a position to offer such definitions, and it avoids mention of significant relations other than gender, such as class, race and sexuality.

Elizabeth Grosz summarises the six philosophical and political limitations of what she describes as 'egalitarian feminism':

1 Sexual equality takes male achievement, values and standards as the norms to which women aspire. At most women can achieve equality only within a system whose overall power remains unrecognised and unchallenged
2 To achieve equality, women's specific needs and interests must be minimised and their commonality with men—their humanness—emphasised
3 Policies and laws codifying women's equal rights have tended to operate as much against women as for them
4 Equality reduces all specificities—including relations of oppression—to oblivion
5 Struggles for equality are easily reduced to social justice and are realisable only in the public realm
6 Even if the two sexes achieve equality, behave in the same way and have the same rights, the higher value accorded to men's activities compared to women's goes unchallenged (Grosz 1988, pp. 93–5).

Thus for Grosz, liberal feminism is fatally limited by its eighteenth century origins and world view. It assumes that women are both fundamentally different from men—primarily because they can have babies—but united by their human status. Relatively minor differences, such as women's alleged physical frailty, have been translated by way of formal and informal regulations—for instance, prohibitions against lifting heavy weights, working overtime and working in certain occupations—into major constraints on women's freedom to work where they choose. Equality between women and men can thus occur through the removal of these restrictions so as to give women choice and freedom. Even though women might be given these choices, however, the foundation of the liberal world view, based in a male-defined social contract, a limited notion of rationality and the distinction between the public and the private realms, means that women are positioned primarily in the realm of the domestic, the lesser, the natural, the bodily and the irrational. Further, the positing of undifferentiated subjects means that critical power relations and dimensions of difference—which derive from race, class and sexuality—are rendered unimportant.

Despite these problems, however, an appeal to liberal principles has been instrumental in securing women's political representation and access to higher education and the professions. Thus, Nancy Fraser, who accepts liberal feminism's limitations and writes from a 1990s feminist landscape preoccupied with differences between women, has called for the re-examination of liberal principles. These principles, she suggests, could overcome our ' failure to connect the cultural politics of identity and difference to the social politics of justice and equality' (1997, p. 99). For Fraser, 'egalitarian feminism's' key insight is its focus on women's social marginalisation and unequal share of resources. In Fraser's view, no feminist politics should lose sight of the liberal goals of equal participation and fair distribution. A rediscovery of liberal feminist principles, she suggests, can counter the immobilising nature of identity politics and the endless fracturing of postmodern feminism in the 1990s (Fraser 1997).

Liberal feminism first appeared in geography in the early 1970s and, despite its limitations, its philosophy has continued to inform progressive descriptive and political work. Despite liberal feminist geographers' apparent preoccupation with the position of women in the discipline and across space, the radical edge of the philosophy, which Nancy Fraser (1997) has recently highlighted, has informed studies of status indexes and social polarisation.

LIBERAL FEMINIST GEOGRAPHY 1: HEAD COUNTS AND REFORMISM

The Women's Liberation Movement and actions by women to achieve progressive institutional reforms and political power was the context in which liberal feminism assumed prominence in Australian geography. Drawing on the views of J.S. Mill, liberal feminists represented the state as a neutral arbiter, pluralistic and open to influence. Though they also recognised that various state institutions—such as the bureaucracy, the judiciary and parliament—were dominated by men and by policies which represented male interests, liberal feminists lobbied for legislation and policies which promoted equality and addressed women's concerns (Watson 1990).

Despite there being few women in positions of legislative power, the role of the feminist bureaucrat, supported by organisations such as the Women's Electoral Lobby, was vitally important during the 1970s in achieving an array of reforms; these reforms included provision of child care, changes in abortion law, education, equal pay and Equal Opportunity legislation. These organisations and changes, and the role of femocrats in positions of relative political power, distinguished this new wave of feminism from its predecessors and confirmed the broader social utility of the liberal feminist position.

Throughout most of the twentieth century, Australia's public–political arena has been profoundly conservative and male-dominated. The statistics for the number of women in parliament and in positions of corporate power, and the wage differentials between men and women, confirm women's secondary status in paid work and politics; but they also indicate that some significant changes have occurred. In the national Parliament, between 1901 and 1992, 396 men were elected to the Senate (the Upper House) but only 34 women, while 833 men but only 16 women were elected to the House of Representatives. In 1992, there were 57 women in the Federal and six State Upper Houses compared with 186 men (ABS 1992). By 1997, the 10 women members of the House of Representatives in 1992 had increased to 13 (or 13% of members), but in the Senate the number of women had declined from 19 in 1992 to 16 in 1997 (or 21% of members) (ABS Women in Australia 1997). Between 1995 and 1998, however, the overall proportion of women in Federal Parliament increased from 14% to 21.4%, double the international average (Walker 1999).

Despite some success in the parliamentary arena, women are still poorly represented in local government and occupied only 19% of Senior Executive Service positions in the Commonwealth public service in 1996. No women were on the management boards of Australia's top 100 companies (ABS 1992), though, in 1998, women occupied 7.6% of all board positions (up from 4% in 1995). In 1996, women occupied 10% of top managerial positions, 16% of second-tier positions and 27% of third-tier positions but only 1.3% of corporate executive directors (Walker 1999).

In the paid workforce, men tend to earn more than employed women and fill the higher level positions; though the gaps are lessening. Thus in 1966, women's full time earnings were 75% of men's. In the aftermath of the 1972 equal pay decision and ongoing agitation, the gap had narrowed to 84% in 1998 (Walker 1998). The academic arena reflects these patterns of continuing, though improving, secondary status for women. Thus the proportion of female academic staff has been rising steadily: from 27% in 1988 to 34% in 1998; but the increase in the proportion of women in positions above senior lecturer has been smaller: from 10% in 1992 to 13% in 1998 (Walker 1999). Such patterns vary across the academic disciplines, with women a lonely presence in the sciences but more numerous in education, health and the arts. Geography, a discipline which strad-dles the arts and sciences, has been slow to change but is now starting to incor-porate more women, but this may relate to difficulties in the discipline. It is relevant then, to look more closely at women in Anglo-American geography.

In 1978, when the British geographer Linda McDowell took up her first acad-emic posting, her attention was quickly drawn to the paucity of women in the hallowed halls of geographical knowledge production and education in Britain. McDowell set about doing a survey not unlike that by Zelinsky five years earlier. McDowell documented the proportion of women undergraduate and graduate students, the number of women teaching in university geography departments and their visibility as writers in the two main geographical journals (McDowell 1979). Her survey, like others conducted in Canada and the USA (see Berman 1977a; Rubin 1979; Momsen 1980; Golledge and Halperin 1983; Lee 1990) and later in Australia (Fahey 1988), confirmed that geography was 'a tradi-tionally gentlemanly profession' (quoted in Johnston and Brack 1983, p. 102). McDowell's figures were startling; they documented the extreme under-repre-sentation of females, and revealed that, in the 1980s, this was worsening. The surveys noted, for example, that, despite the wider political context of Women's Liberation in the 1970s, in the 1940s, over 30% of geographers obtaining their first academic positions in British geography departments were female compared with 11.4% in 1982 (Johnston and Brack 1983, p. 110). Table 1.1 summarises some of the results from surveys published on British and North American geography.

Country	Students				Staff	
	Majors	Masters	PhD	Universities and Colleges	Professors	Members of National Associations
United Kingdom						
1978	42	29	31	52	7	–
1982	45	–	25	11	1	–
USA						
1970	–	12	12	6	13	12
1976	–	–	26	14	–	–
1979	–	–	4	–	–	–
1980	–	25	18	10	–	–
1989	–	–	27	8	–	10
Canada						
1970	–	5	6	–	–	–
1978–79	39	14	6	7	–	–

Table 1.1 Women as a proportion in British and North American geography, 1970–89
Source: Mc Dowell 1979; Johnston & Brack 1982; Momson 1980; Monk 1984; Rubin 1979; Zelinsky 1973b; Zelinsky, Monk & Hanson 1982.

These surveys show that the situation of women in contemporary Anglo-American geography between 1972 and 1992 was as follows:

● while women comprise nearly 50% of undergraduate students in the discipline, the proportion who go on to higher degree study and thence into academic positions declines sharply

● the few women who enter the academic system tend to be in part-time, untenured and junior positions

● women are less visible in the discipline's public arenas, and this includes their lack of presence at conferences and in publications. Thus, for example, while 16% of those who attended the 1983 Institute of Australian Geographers Conference were women, only 2% of those who gave papers were women (Howe 1984, p. 151)

● similarly, women tend not to hold senior positions in the associations connected with the discipline; they are less likely than men to be on the editorial boards of the leading journals or on the executives of the professional associations

● despite claims that the situation is improving, the evidence is inconclusive. In the USA, Janice Monk notes the number of female PhD holders increased from 7% in 1972 to 18% in 1992, and, in the same period, the proportion of women in the Association of American Geographers rose from 13% to 27% (Monk 1994). In Australia, the situation is less favourable.

Table 1.2 provides some data on the sex of authors in two Australian journals.

	Australian Geographical Studies						Australian Geographer					
	1973–83		1984–94		1995–98		1973–83		1984–94		1995-98	
	No.	%	No.	%	No.	%	No.	%	No.	%	No.	%
Female	21	11	46	10	32	24	20	9	36	6	18	15
Male	133	67	284	63	75	57	130	59	422	71	68	58
Unstated/ Unknown	45	23	123	27	25	19	71	32	140	23	31	27

Table 1.2   The sex of authors in two Australian journals, 1973–98
Source: Johnson 1985 and the author's research

While no firm conclusions can be drawn from table 1.2, given the proportions where gender is unknown, it is clear that, in the two key geographical journals in Australia, male authors published at a much higher rate than female authors between 1973 and 1998; and that it seems likely that the rate of female authorship may have increased in the three years between 1995 and 1998, compared with the two previous decades.[2]

Table 1.3 shows the number of female academics in Australian geography departments. Their poor representation of 21.5% is likely to worsen, given the actual demise and merger of many geography departments.

University	Male	Female	Unknown	Total
ADFA*	13	3	4	20
Adelaide	15	6		21
ANU**	9	2		11
James Cook	23	6	1	30
Curtin	3	0		3
Flinders	12	3	1	16
Macquarie	15	9	7	31
Monash	19	8	2	29
Newcastle	14	8	1	23
Sydney	16	2		18
Uni Melbourne	15	6	1	22
Uni New England	19	3		22
Uni New South Wales	15	4	1	20
Uni Tasmania	10	7	1	18
Uni Queensland	18	1		19
Uni South Aust	24	3		27
Uni West'n Aust	14	4	1	19
Wollongong	17	5		22
Total	271 (73%)	80 (21.5%)	20 (5.4%)	371

Includes geography, environmental science, geographic information systems, geomatics and planning academic and research staff who are associated with geography.
* Australian Defence Force Academy
** Australian National University

Table 1.3   The sex of geography staff in Australian universities, 1999
Source: University Handbooks and websites

Within the Institute of Australian Geographers, which is the professional organisation of the discipline, there is also an imbalance of men and women; though again, there has been a recent trend to increase the number of women on the Council.

The compilation, contextualisation and interpretation of these figures gives some insight into the nature of geography in the countries concerned, and the liberal feminist politics which inspired their collection.

For Zelinsky, the 1970s pattern was 'systematic and structural' rather than arising from chance or from the actions of some individuals. The explanation lay in 'the sexist structure of the larger social system' which emerged from 'our long agricultural and pre-agricultural past, with its rigid sexual division of economic roles' (Zelinsky 1973a, p. 102). To this explanation should be added current practices, since '(t)he immediate causes are the institutional rules, traditions, and biases, usually unspoken and unwritten, of the organizations in which they are trained and employed' (Zelinsky 1973a, p. 104). As a result, 'organized action and agitation (in the form of) education, propaganda, political pressure and even legal suites' will necessarily occur and, by implication, can be effective.

A number of geographers, including Mildred Berman (1977a), Linda McDowell (at least in 1979), Barbara Rubin (1979), Janet Momsen (1980), R.J. Johnston and E.V. Brack (1983) and David Lee (1990), see the issue of women in geography as that of their absence from higher level study, from senior faculty positions, from professional associations, from editorial boards and from authorial visibility. The solution then becomes the presence of women at those sites where geographical knowledge is created, circulated and legitimated. Such a presence would both enrich the discipline and remove, in this arena, the injustices associated with the marginalisation of over half the population. This view assumes that, though women and men may be socialised into different roles, they have the same capacities and should, therefore, be equally represented in geography.

This view contrasts with Wolf Roder's (1977). In an exchange following Zelinsky's 1973 articles, Roder argues that there are 'real' biological differences between women and men. These bodily and psychological differences, he asserted, underlay women's inability to conceptualise spatial relations and meant that men and women had different geographical abilities. Consequently, 'we shall never be able to recruit fifty percent of geographers from the female gender' (Roder 1977, p. 398). In reply, Zelinsky (1977) and Berman (1977b) question this biological determinism and affirm their position on socialisation and sex roles. In this formulation, maleness and femaleness are roles which are learnt and are thus open to change. In liberal feminist theory, sex roles may vary in their manifestations but they are also grounded in the distinction between the public and private realm and relatively fixed across an individual's life. Sex roles, according to these writers, are also not modified by class, race, ethnicity or sexuality. Consequently, gender difference remains the sole focus of those conducting head counts of women in liberal feminist geography, so that the number and proportion of geographers who are from non-Anglo-Celtic background, who are gay or who come from working-class backgrounds, remains undefined.

This focus on gender undifferentiated by class, race or sexuality, which Pateman and others have noted, co-exists with the assumption in such head counts that (white) women's sex roles assume domestic responsibilities but that (white) men's do not. This assumption underlies the early writings of Wollstonecraft, Mill and Taylor; it also permeates Mildred Berman's 1975 survey of members of the Association of American Geographers. She writes of the survey design: 'There were some differences in the questions asked of the men. Items asking about the age of the youngest child and reasons for not continuing studies toward the doctorate were omitted on the assumption that child care was not likely to be a hindrance to the professional career for men' (Berman 1977b, p. 71).

Berman subsequently notes that her research confirmed this assumption. She thus elevates as historical truth the view that for men child care is not an issue in their career whereas for women their maternal and parental responsibilities are vital in shaping or hindering their careers.

Measures to achieve the requisite gender balance reflect this and other liberal feminist assumptions in geography departments, and in research and teaching. Thus, for instance, Zelinsky urges individual women to take action on their own behalf, using Equal Employment Opportunity (EEO) and affirmative action legislative frameworks. Berman echoes this when she observes: 'The legislation which has been written and passed must be enforced and for this women denied equal accessibility to professional opportunities must articulate their dissatisfaction and file their grievances with the appropriate federal agencies' (Berman 1977a, p. 9). There are also calls for financial support for women to do graduate study (Lee 1990), for 'fairer hiring practices' (Beckwith 1977; Berman 1977a), for child care and for special courses about women (Golledge and Halperin 1983). This resonates with liberal feminists in the eighteenth and nineteenth centuries when they lobbied for the vote and for removing restrictions on women's entry to parliament and higher education. The present agenda goes further however, in tackling directly the particular problems women face with child-care responsibilities, lower incomes and discriminatory hiring practices in universities. In addition, there is a recognition that existing curricula do not explicitly consider women and that this needs to change for equality to occur. In a liberal feminist politics, it is individuals, departments and professional associations who are to enact these pro-woman changes.

These head counts, studies and practical suggestions were limited by their liberal feminist framework. But they were also the first manifestations of feminism in the discipline, appearing in the 1970s and 1980s. The assumptions informing these studies are that:

● men and women are individuals united by their human capacities and potentialities but also divided by their biology and sex role socialisation
● the subordinate position of women, while historical and structural in origin, is subject to change
● the most necessary changes are for women to attain a presence within the discipline equal to that of men. This presence will occur if barriers to women's access to the visible and higher echelons of the discipline are removed and compensatory measures put in place which recognise women's domestic responsibilities
● such changes will come about through women organising and forcing men to act

- effective change can be achieved in existing departments and institutions through legislative frameworks.

First articulated over twenty years ago, these analyses and strategies still exist in feminist geography. Thus, in 1990, Linda McDowell argued that British universities were 'patriarchal organisations' in which there was verbal and visual abuse of women, sexual harassment, an oppressive sexual division of labour and a body of knowledge which was generally insensitive to gender. In passing, she notes that under-representation was a problem for blacks as well as women but does not pursue this further. In response and within a liberal feminist framework, McDowell suggests a raft of policies and organisational changes. Thus she proposes an ongoing statistical record of 'minority' groups, and the setting of targets 'for achieving a more representative staff and student body', policies to restrain staff and students from engaging in verbal and physical harassment as well as formal and informal organisations to deal with cases, a greater range of courses focusing on ethnic and gender diversity, training in non-sexist recruitment and selection procedures, and policies to enhance the structural power of often relatively junior women academics (McDowell 1990). Significantly, extending beyond liberal feminist assumptions, such changes in existing structures and practices were not to be precipitated solely by the women who suffered disadvantage. Instead, McDowell argues:

> it is men who must choose to change. The men who currently hold power within academia will be those who must introduce and support policies to widen access to a diverse range of students, staff and areas of knowledge. The men who currently hold almost all the teaching jobs in British higher education must choose to support the changes and allow access to those who are not in their own mould. It is also men who have to recognise that currently they use their power in various ways, some explicit, others inadvertently, to oppress and disempower others. Men have much to lose, but much to gain; the future of higher education is in their hands (McDowell 1990, p. 331).

It is to men as power holders, then, that this more radical feminist analysis points; but aspects of the liberal feminist agenda remain. These include assumptions about equality, the need for change through alterations to existing policies and practices, the focus on gender difference within a basic humanism and the importance of attitudes. However, McDowell highlights the radical dimensions of liberal feminism which so disturbed those who resisted demands for women's suffrage and higher education in the nineteenth century. In her calls for equity and justice, she affirms the progressive core of liberal feminist philosophy as Nancy Fraser has recently described it.

In presenting her case, McDowell points to the groups which have made such calls feasible and achievable: notably in the United Kingdom, the Gender and Geography Group within the Institute of British Geographers. Canada and the USA have long had 'Committees on the Status of Women'. The British Gender and Geography Group is the model for similar groups in Australia (1988) and New Zealand (1992), and within the International Geographical Union (1988). Geography and Gender Groups have been instrumental in networking female geographers, providing information on jobs, setting up meetings and circulating lists of new publications in regular newsletters, and facilitating discussion as well as research and publication activity. These groups have begun the process of countering the gentleman's club so dominant in English-speaking universities, while also generating an alternative cooperative and collaborative way of working to change the culture of academic institutions.

The Australian group, formed in 1988, has been visible at every national conference: it has created an active and supportive network, publishing blocs of articles in the major Australian journals (see *Australian Geographical Studies* 1988 and *The Australian Geographer* 1998), generated subgroups which have acted as localised supportive environments, discussion forums, critical sounding boards for new work, achieved funding and publication, supported mini-conferences and encouraged its feminist graduate students. Like its counterparts elsewhere, the Gender and Geography Group in Australia has also been instrumental in raising the issue of equal opportunity and affirmative action within the Council of the Institute of Australian Geographers (at Newcastle in 1995, though without any legislative outcome). Over the 1990s, the number of women in senior university and professional positions in Australia and New Zealand has grown as has their public profile through publication and presence at conferences. However, despite these significant achievements, such developments have not yet significantly altered the demography of the discipline (see figures 1.1–1.3) nor the discourse of geography.

LIBERAL FEMINIST GEOGRAPHY 2: MAPPING THE STATUS OF WOMEN

Women's absence from geography departments and from other sites of knowledge creation and circulation has thus been questioned. There has also been a systematic appraisal of the content of the discipline, as a subject taught to students and a body of research and publication. Thus in 1976, in a session on 'Discovering women's place' at a meeting of the United States National Council for Geographic Education, delegates noted that geography was deficient in subject matter in dealing with women. It was pointed out that this had led to distorted theories, impoverished regional accounts and courses which portrayed women—if at all—as passive rather than as active agents of social or landscape change. In response, two strategies were put forward: one was to inject a feminist perspective into the regular syllabus through supplementing texts and lectures, and the second was to create new courses focusing on women (Lloyd and Rengert 1977). Twenty years later in Australia, only the first strategy has been adopted and only sporadically.

Thus, in 1987, Stephanie Fahey surveyed members of the Institute of Australian Geographers (IAG) on the importance of gender issues in their teaching. She found that three quarters of the 25% of IAG members who responded saw gender as relevant to their courses, primarily in the areas of human, Third World, physical, economic and population geography. However, these respondents included gender only as a piecemeal addition, as a lecture, seminar or essay topic on women, rather than through restructuring courses or through the systematic teaching of feminist geography (Fahey 1988, see also Deatherage-Newsom 1978; Lee 1978; Rengert and Monk 1980; Monk 1984).

A similar approach characterised research and publication. Thus in 1982, Janice Monk joined Susan Hanson to observe that 'through omi[tting to] consider...women, most geographic research has in effect been...sexist' (Monk and Hanson 1982, p. 11). They isolated sexist bias in the content, methods and purposes of geographic research. On content, they discussed studies of migration, development, suburbanisation and urban travel which did not consider female as well as male experience. They also found that the choice of research areas was biased so that women's activities were deemed insignificant, and that research avoided topics which directly addressed women's lives such as child care and unpaid work. Where there was explicit writing on women, it tended to assume traditional gender roles (Monk and Hanson 1982, pp. 14–17). So, too, in methods. The choice of variables and respondents generalised from male subjects: for example, in the use of the husband or the '(male) household head' to refer to the whole household. Similarly, Monk and Hanson argued that in interviewing techniques research practices further entrenched sexist assumptions in the production of geographical knowledge (Monk and Hanson 1982, pp. 17–19). Feminist geographers have also noted the continued marginalisation of women in other areas of recent geographical concern, such as localities and gentrification (see, for example, Bowlby 1986; Warde 1991).

However, calls by feminists to include women within a range of geography sub-disciplines has produced a remarkable efflorescence of studies of 'women and...transport, migration, Third World Development, agriculture, recreation and tourism'. For example, in a 1977 study on journeys to work, it was noted that research in the field rarely disaggregated by sex, concentrating instead on the 'heads of households' in their data collection and analysis. Julia Erickson proceeded to examine 5000 black and white women aged from thirty to forty-four to observe how married women travelled shorter distances than unmarried women, and how white women journeyed further and longer than black women to get to paid work. Erickson related these patterns to levels of suburbanisation, car ownership or dependence on public transport and domestic responsibilities (Ericksen 1977). Similar studies confirmed these patterns for the USA (see, for example, Lopata 1980) and for Australia (Howe and O'Connor 1982), while the Australian study added the important dimension of the gendered labour market to its explanation. These studies were vital correctives to earlier work and indicated a discipline newly sensitised to gender. Such work was also often accompanied by calls to alleviate the patterns of disadvantage it revealed. In these respects, this work represented many of the strengths of the liberal feminist position. However, it also illustrated some of the narrowness of liberal feminist geography: its limited focus on gender difference, and its emphasis on description and reformist solutions.

To gain a deeper sense of the strengths as well as the limitations of the 'putting women in' strategy, it is relevant to look at overseas studies which aim to include women through mapping the place of women.

The Status of Women in Australia

In 1983 Mary Ellen Mazey and David Lee wrote, under the auspices of the Association of American Geographers, the 'first published geography monograph on feminist issues' (Mazey and Lee 1983, p. v). In this work, they asked what they saw as fundamental geographical questions of women: where are they? how and why do they move in space? what do they think of the environment? and how do they make their mark upon it? Answers to these questions were somewhat schematic and drew on pre-existing research. In their use of descriptive work done over the 1970s, and in their concern for an inclusive geography in which women's sex roles were both documented and challenged, I would argue that Mazey and Lee constituted their project as a liberal feminist geography.

Mazey and Lee's book, *Her Space. Her Place*, begins by considering sex ratios over time and space before looking at women's rights and status globally and in the USA. They map social and economic indicators and disparities across the USA before examining the fate of the Equal Rights Amendment and regional variations in abortion and the use of contraception. A more recent atlas,

Women in the World (Seager and Olson 1986), extends this mapping technique to detail global patterns of marriage, mothering, work, access to resources, welfare, authority as well as 'Body Politics' and the politics of assembling the statistics for what Seager and Olson call 'Mapping the Patriarchy'. Both texts provide powerful compilations of information and visual statements on the place of women in various countries; and both affirm the ongoing utility of the liberal feminist project and the power of visually representing gender inequality.

Australia appears in both books as a unified whole, one which rarely diverges from Euro-American patterns in the measures used, an exception being pro-woman legislative changes which have been determined at a State level (such as suffrage and abortion laws). In order to establish the utility of statistical indicators for documenting the extent and spatial variability of women's inequality, and to highlight some of the limitations of such a strategy, I replicated Mazey and Lee's US study on female status. The study is of statistical divisions in Victoria and Melbourne and it examines high female income levels (over $50 000 p.a.), the proportion of women with tertiary qualifications and those in high status professions in order to give an overall Index of Women's Status across these statistical divisions.

The patterns in figure 1.2 are instructive. The highest status women tend to live in the centre and eastern parts of Melbourne, and in a number of coastal and rural areas. One explanation of these concentrations, which include the Bellarine Peninsula, Ballarat, Bendigo, the north east and south west of Victoria, may be the presence of local universities; these institutions welcome mature aged students and may thereby inflate the numbers of tertiary educated women in these localities. More significantly, though, are those areas of low income, and low levels of education and employment. The La Trobe Valley to the east of Melbourne shows a particular concentration of low status women; this may be related to the nature of the coal mining and power generating industry which dominates this region, an industry which has shed many (male, blue collar) workers and which exists in an area where female employment has only recently begun to rise significantly in occupations such as hospitality and retailing where education and income levels are low.

Figure 1.3 provides more detail on women in Melbourne. It shows that both low and high status women are concentrated in wedge-like patterns across the city. As in figure 1.2, the inner and near eastern suburbs and a few bayside suburbs accommodate the highest status women; this pattern recurs for high status men (see figure 1.7). Whether the pattern for women arises from their particular role as gentrifiers in the inner city, or from their class connections to middle-class and professional men in a city long divided into east and west by

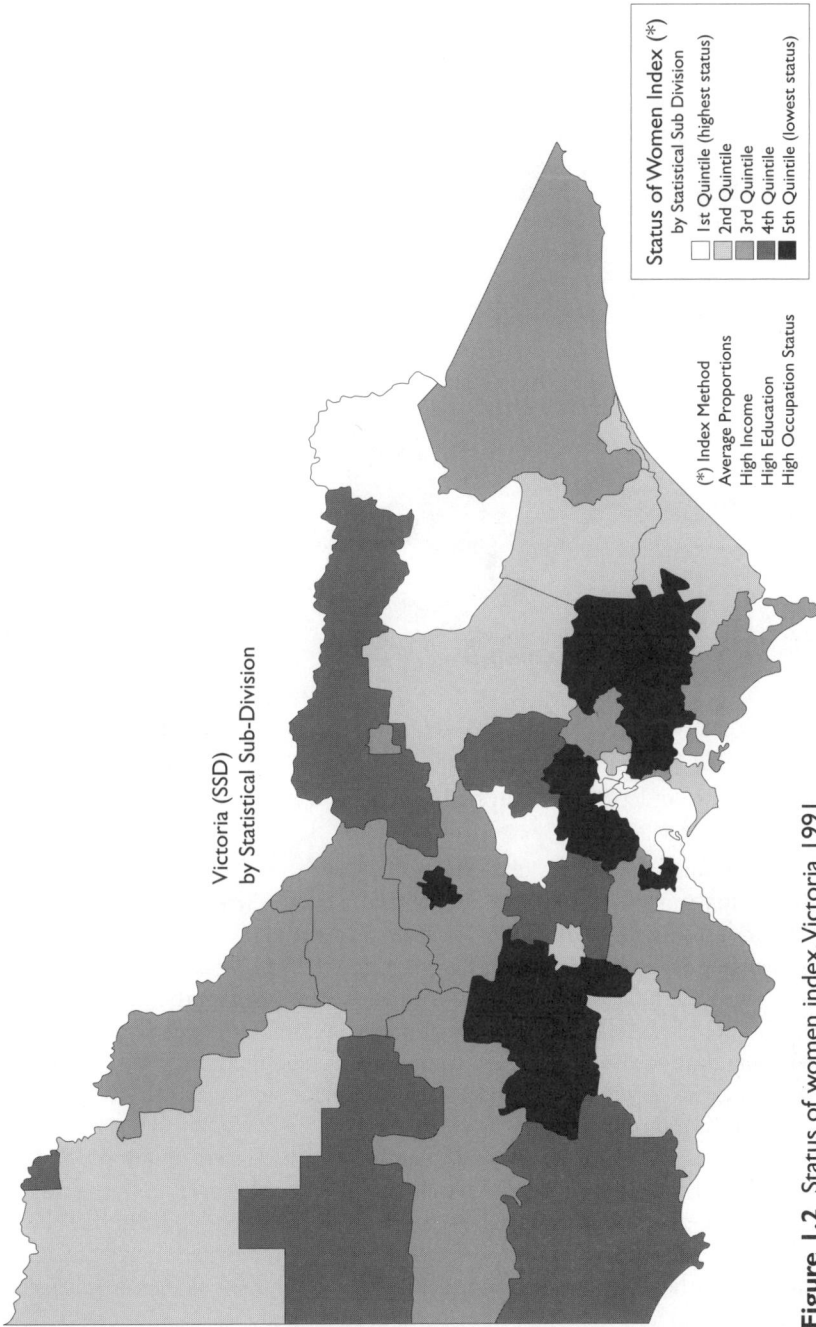

Figure 1.2 Status of women index, Victoria 1991
Source: ABS 1991 Census. Map by DivaData

Victoria (SSD)
by Statistical Sub-Division

Status of Women Index (*)
by Statistical Sub Division

1st Quintile (highest status)
2nd Quintile
3rd Quintile
4th Quintile
5th Quintile (lowest status)

(*) Index Method
Average Proportions
High Income
High Education
High Occupation Status

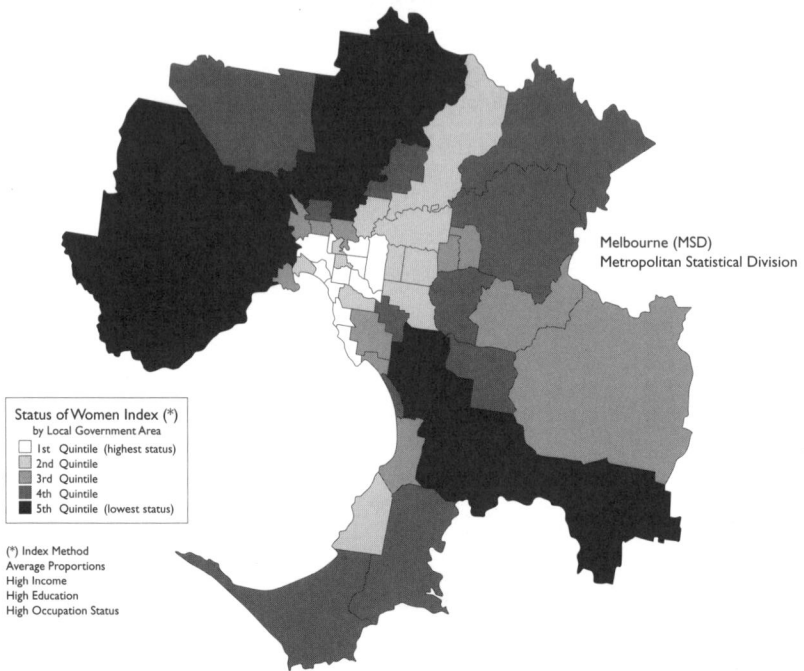

Figure 1.3 Status of women index, Melbourne 1991
Source: ABS 1991 Census. Map by DivaData

class, can partly be answered by dissagregating the various elements which make up the Status of Women Index. Thus, as figures 1.4–1.7 indicate, women in these areas are independently in high status occupations and have high levels of education and income.

These detailed maps for Melbourne, show that:

- individual measures of those women who are well qualified, have high incomes and who work in professional and managerial occupations show somewhat different patterns. There is a divergence, especially between the inner city as the abode of highly qualified women, the north eastern suburbs as the site of high status professionals, and the middle ring eastern and western suburban location of high income earners. Such patterns invite different explanations as well as raise questions about the oft-assumed equivalence of these measures of high status
- when combined into a 'Status of Women Index', the eastern and north-eastern suburbs of Melbourne emerge as the sites for privileged women to live, with the outer ring of suburbs—with the exception of Werribee and Sherbrooke— accommodating those women of lowest status. Such a pattern raises questions about why there are concentrations of high and low status women in particular locations and how these may have changed over time
- when compared with a 'Status of Men Index', a pattern emerges of a class— rather than a gender-differentiated city: the location of high status men pretty well matches that of women.

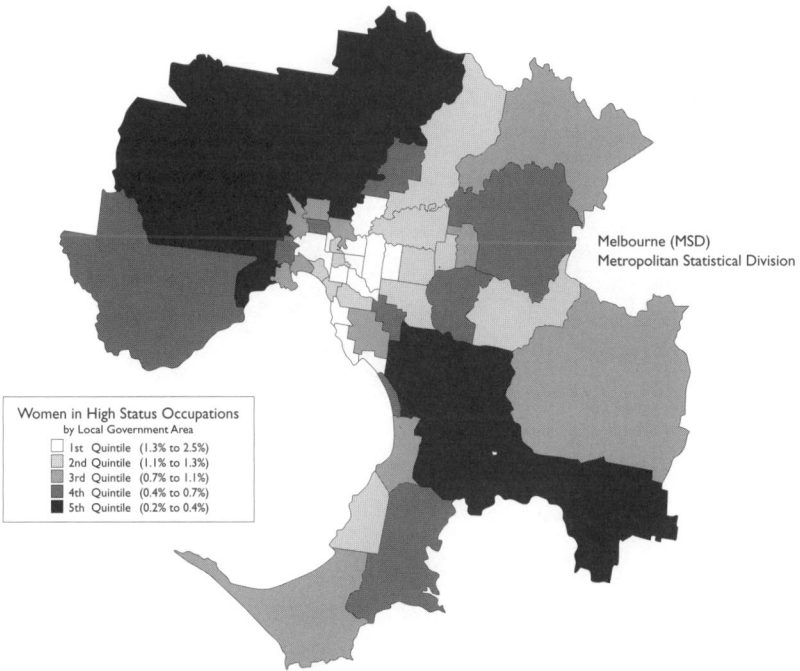

Figure 1.4 Women in high status occupations, Melbourne 1991
Source: ABS 1991 Census. Map by DivaData

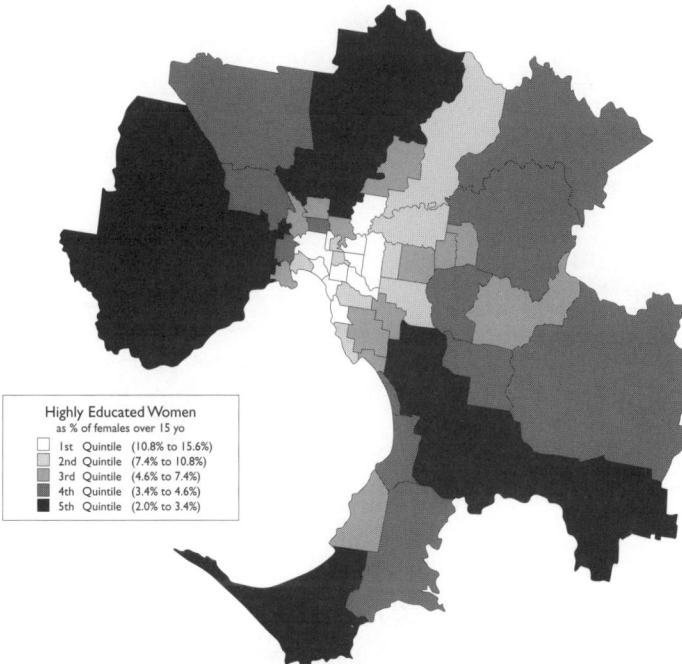

Figure 1.5 Highly educated women, Melbourne 1991
Source: ABS 1991 Census. Map by DivaData

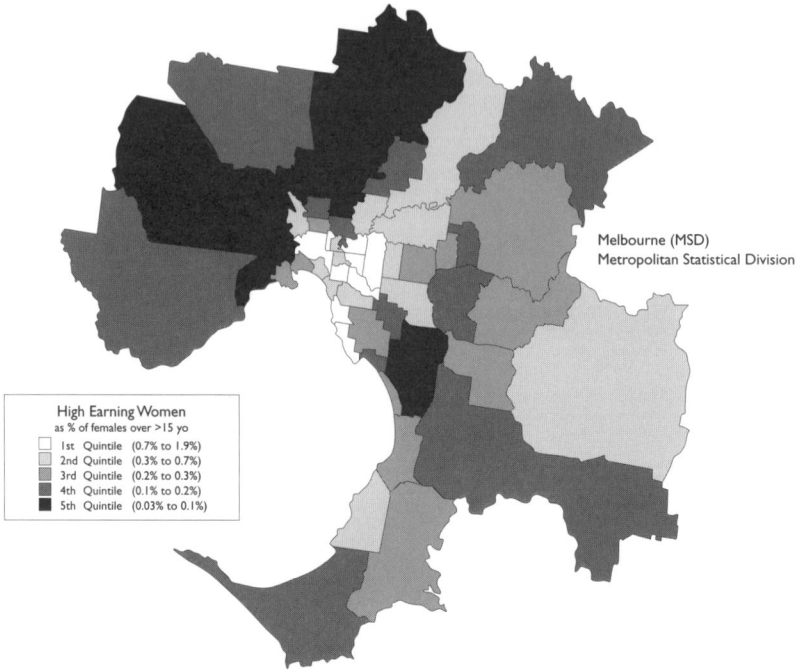

Figure 1.6 High earning women, Melbourne 1991
Source: ABS 1991 Census. Map by DivaData

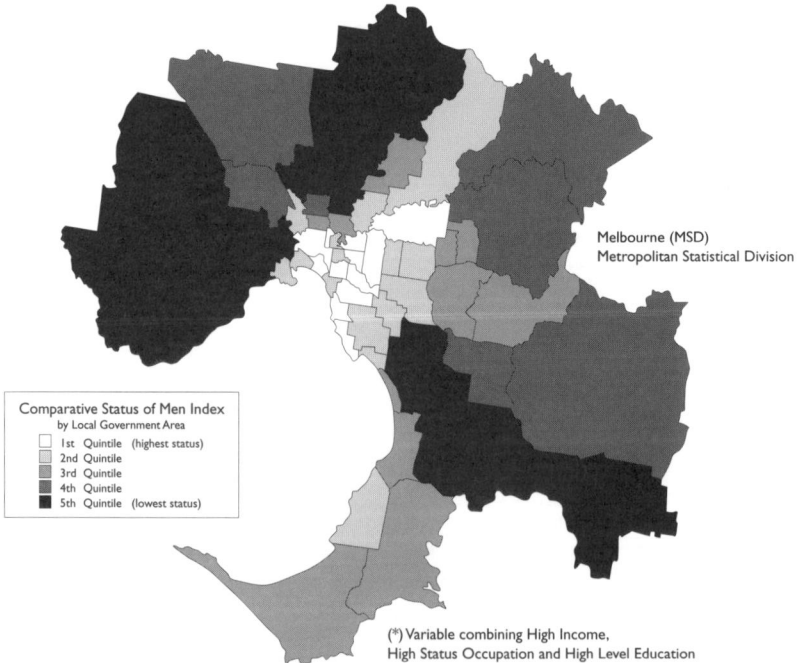

Figure 1.7 Status of men index, Melbourne 1991
Source: ABS 1991 Census. Map by DivaData

Despite the insights from, and the useful questions raised by these mapping exercises, such maps have limitations:

- The maps are very much descriptive devices. They may highlight general patterns but they tend to raise more questions than they answer. They map where women are, and attempt to provide visual displays of patterns of inequality, but they do not tell us why such patterns occur
- The main variable is gender, with a subsidiary one related to a general notion of class measured through levels of education, occupation and income. As a result, cities like Melbourne do indeed appear gendered but, depending on the technique used, such a differentiation will primarily be in terms of gender or income, education and occupation. 'Inequality' therefore becomes, in part, a function of the variables selected
- The gendering of urban space, while a lived reality for many in the city, is obscured rather than clarified in such a mapping exercise. The scale of such maps is large, so that it does not reveal many localised and invisible aspects of gendered spaces. Mapping is a visual tool, one which is also dependent on statistical data. As a result there are very real limitations in what mapping can show
- The complex nature of spatial and gender relations is obscured by this technique. For example, the role of ethnicity and household form in the creation of these patterns is not at all clear; household form includes the number and age of children, number of income earners, part-time and full-time workers, nuclear, blended or single person household. The role of spatial mobility and the quality of different urban environments for the realisation of high status or the management of low status is not indicated in such an exercise. The technique is flawed by the limited number of variables which can be included, by scale and dependence on the quantifiable and the visual.

From Status Index to Social Polarisation

While it is no longer fashionable to address the issue of gender equality in such descriptive and cartographical terms, the question of social differentiation remains of great concern to geographers. In particular, the question of women's unequal position within a restructuring economy has been addressed by literature on social polarisation, a literature which, in many ways, newly articulates the liberal feminist concern with social divergence.

Thus when surveying major socio-economic changes in contemporary Australia, geographers, economists and sociologists have noted that the sectoral shifts from manufacturing to services, and changing labour relations have led to an increase in social and spatial polarisation. Thus research on Adelaide by Forster, and by Baum and Hassan, concluded that certain areas within the city have sustained high levels of unemployment, accommodate large populations on low incomes, have a falling proportion of women in the workforce, and high numbers of public housing tenants and single parent households. In contrast, the more affluent areas have not been as affected by high levels of unemployment, by falling numbers of women in the workforce or by the growth in low income households (Baum and Hassan 1993; Forster 1986). Frank Stilwell made similar observations for Sydney in 1989, while the Australian Institute of Family Studies noted growing disparities in income in both Sydney and Melbourne between 1986 and 1991 (Stilwell 1989; Burbidge 1994). Such polarisation has occurred across whole cities and within them (Marcuse 1995; Pinch 1993): for example, across Western and Northern Sydney, but also in particular parts of the Western suburbs. As Burbidge and Winter have noted, these patterns cannot be explained simply by economic restructuring; they must also take into account changing household forms, the location of public housing and the specifics of suburban migration (Burbidge and Winter 1996; Badcock 1997).

In one example of feminist geographers coming together to explore an issue of importance and to use their collective wisdom to secure research funding and a publication, a number of Melbourne women, under the auspices of Katherine Gibson and Ruth Fincher, gained research monies from the Australian Housing and Research Institute to explore the gendering of social polarisation (Gibson et al 1996). While not working explicitly within a liberal feminist framework, the project was committed to women as a social group and to social justice. Team members explicitly located themselves within postmodern feminism (see chapter 4) yet the work produced a theoretical and practical set of tensions between these liberal, feminist principles and postmodern theories of social difference. As Katherine Gibson notes in her opening essay, discussions around social polarisation and the politics of difference were 'discourses in collision' (Gibson 1996): one discourse tending towards quantified, class-based activism and social collectivities and the other towards social fragmentation, qualitative methodologies and a focus on a range of identities and places. Such a sentiment echoes Nancy Fraser's observations on the tensions for larger social collectivities between a concentration on identity and micro-differences and questions of social inequality and justice. The eight contributors to the project differed in their resolutions of these tensions, though in Jenny Cameron's work an insightful postscript can be added to the Status of Women Index for Melbourne in 1991.

In her study of women's part-time work, Cameron elevates the importance of women's motivations as well as their household situations in shaping their identities and location in any socially polarised city. Their household situations are differentiated by children, race, sexuality and ethnicity (Cameron 1996). The patterns her analysis generates, as she examines the relationship between women as part-time workers and their household situation, is quite unlike that which emerged from mapping high income or high status women. As with existing social polarisation studies, the focus of the Status of Women Index is on women as individuals not as members of households. The contrast in maps of women's part-time employment and high income women confirms Cameron's argument that women in part-time employment are often there because their husbands are high income earners, and also because these households are buying housing in areas where there is an abundance of part-time work for women. The reasons for the patterns therefore go well beyond a (liberal) unidimensional view of women as individuals (Cameron 1996).

Cameron concludes that the preoccupation in the literature on social polarisation with the economic aspects at the expense of the socio-cultural, the global at the expense of the local and with class at the expense of other social markers, limits its arguments. But in affirming the need to focus on women as a social group with particular needs and disadvantages, Cameron continues to use aspects of the liberal feminist tradition while also recognising its inherent limitations.

CRITICAL REFLECTIONS

Liberal feminism is a philosophy which has a number of strengths; these have informed a range of significant political achievements over the last two centuries, and its view of justice and equality retains its relevance. With an emphasis on visibility within the academy—in employment numbers, in publication, in research, in recognition—in things that matter—liberal feminist geographers engage with a number of issues of immediate and real concern to women, and have goals and strategies for their realisation. Thus liberal feminism has an agenda of political reforms to remove barriers and to secure changes to enhance the presence and success of women through the provision of, for example, child care, anti-discrimination legislation and equal opportunity commitments. Over the years, there have been measurable successes, especially in relation to the attainment of political changes. These include the vote for women, pro-woman legislation and child care. Such successes have been achieved through existing channels of power, and through the creation of new political groupings; these new groups, such as women's networks, caucuses and organisations, are often cooperative and function in a way quite different from existing structures.

Within geography, liberal feminist geographers have successfully critiqued the content, method and approaches of the discipline, documented the parlous position of female geographers in the academy, and begun a process of making women visible in the discipline and in geographical space. Visibility is therefore a strategy which has had a real and vital impact on the discipline. Without this lobbying, women's position and that of feminist geography would be far worse than it is today.

Despite these many achievements, liberal feminism's inherent philosophical and political limitations assume specific forms in geography. The philosophy is built on a nineteenth-century conception of gender which, in the name of a common humanity, assumes a sexual division of labour, heterosexuality and an occlusion of class, ethnic and racial difference. In its political agenda, liberal feminism strives for equality, but the standard against which such equality is measured is the position held by men. Thus, in the quest for liberal reform, the gender order is not fundamentally challenged, while its masculinist foundation remains unquestioned and unaltered.

Within feminist geography, it has been assumed that visibility would effect a revolution: that critiques of the discipline, a presence of women in professional organisations, in writing, on editorial boards and in research and teaching prac-tice—as well as the mapping of women's inferior position—would fundamentally alter the discipline. And in many ways it has. However, not only has the number of female geographers not appreciably increased, but their visibility in Australian geographical journals, for instance, has not risen markedly. Similarly, while gender is now a topic in many courses, its inclusion is, at best, small and tokenistic and has not achieved a thoroughgoing change in the way in which geography has been conceptualised or taught. In short, visibility has not been effective, nor has it meant greater power, autonomy or fundamental change in the discipline.

Consequently, the liberal feminist position was supplanted in the 1980s in Australia by socialist feminism (see chapter 2); it was somewhat out of favour in the 1990s but has more recently been enlivened by radical, postmodern and post-colonial feminisms. However, I would argue for the absolute necessity of contin-uing liberal feminist strategies. It was the liberal feminist concern with head counts which encouraged my own study of head counts of women in Australian universities and in geographical journals. Before I carried out this study, I had assumed that the position of women in these areas was improving, that women had made great strides, but this is not the case. Compiling a Status of Women Index has been a useful exercise in raising questions about the socio-spatial differ-entiation of places like Victoria and Melbourne which have not been suggested before. Such an index has also provided a useful sounding board for arguments around social polarisation in the city. Thus the perspective is not without its political strengths and should not be dismissed because of its limitations.

Notes

1 Harriet Taylor was a long-term collaborator, companion and, eventually, the wife of John Stuart Mill. He frequently acknowledged her assistance in formulating and in writing his ideas. She also wrote significant tracts in her own right, though she is far less well known as a key liberal theorist. Feminists such as Spender (1982), Okin (1979) and Eisenstein

(1981) see *The Subjection of Women* as a collaborative piece but also as less radical than Taylor's 'Enfranchisement of Women' which argued that women be able to enter all professions and levels of education for their own sakes—not just to provide a choice other than marriage. In contrast, Rosemarie Tong (1989, p. 17) suggests that there is general agreement that Mill and Taylor collaborated on the 1832 'Early essays on marriage and divorce', that Harriet Taylor was the primary author of 'The enfranchisement of women' (1851) and that Mill was the main author of *The Subjection of Women* (1869). Despite the obvious historical disagreement, here I will regard *The Subjection of Women* as a collaborative piece.

2   Nevertheless, there remains an extraordinary imbalance; as Table 1.2 shows, in the last 14 years, men have published over 60% of all articles in the main geographical journals. Further, when Table 1.3 is considered, the proportion of women in Australian university geography departments of 20% is well above their usual presence in the publishing arena.

2

Socialist Feminist Geography

If the 1970s were dominated by a liberal feminist appraisal of geography, by the early 1980s, some of its political and theoretical limitations were being noted by a new generation of women emerging from Marxist or 'radical geography'. The Canadian Suzanne Mackenzie offered a critique of the pre-existing work and, as a socialist feminist geographer, pointed to new research directions (see also Foord 1980; Brownhill 1984).

> To date, the primary object of the 'geography of women' has been the empirical measurement of women's spatial perception and activities...(but) If we are truly interested in a 'geography of future equality' we must not be content with the mere documentation of ahistorical spatial restrictions, nor with partial and static explanations that assume a continuity in women's position. We must begin with the daily changes in the social relations of productive and reproductive work which create, reflect and unremittingly recreate and reflect anew women's relation to the environment (Mackenzie 1980, p. 49).

The possibilities which historical materialism offers of a more structural, explanatory, rigorous, and liberating analysis of women's position has been articulated for at least a century. Within geography, however, it was not until the early 1970s, in the pages of the new journal *Antipode: A Radical Journal of Geography* that this connection was first made (see Burnett 1973; Bruegel 1973; Hayford 1974). The social and historical context of this work was the 1960s Women's Liberation Movement and the ascendancy of Marxism as a political philosophy, informing Third World liberatory struggles, the Black Power Movement in the United States and an upsurge in working-class radicalism across Europe and Australasia. However, it was ten years before researchers within radical geography circles—both men and women—became committed to applying Marxism to geographical problems and published substantive work in socialist feminist geography.

The theoretical elements which comprise socialist feminist geography derive from Karl Marx and Friedrich Engels' conceptualisation of capitalism as a set of productive and reproductive relations. While only Engels specifically related women to this system, and arguments rage as to whether Marx's analysis of capitalism was essentially gender blind or profoundly sexist, socialist feminist geographers have used this general framework—and its various modifications by feminists over the 1970s and 80s—to analyse cities, regions and localities. In this chapter, I will firstly, critically outline the theoretical foundations of socialist feminism before looking more closely at how it has been used in the geographical study of women in capitalist societies. Finally, I will present a study of economic restructuring in Geelong, a regional centre in Victoria, from a socialist feminist perspective.

SOCIALIST FEMINISM

The socialist tradition before Karl Marx and Friedrich Engels, takes different forms. For example, eighteenth- and nineteenth-century European Utopian Socialists had gender oppression under capitalism as a central concern. The rise of industry, male-dominated trade unionism and formally organised Leftist political organisations, saw the marginalisation of women. In socialist theorising, it led to a focus on paid labour and class relations (Taylor 1983). Some feminists argue that this marginalisation derives from Marxist theory itself; hence the inclusion of women into Leftist theory, the very creation of socialist feminism, has long been a difficult task, one limited by the masculinist nature and discursive dominance of Marxism (see Campioni and Gross 1983; Johnson 1990a). Some feminists have reconceptualised Marxism, while others have abandoned it. Within geography, some continue to work within a historical materialist framework and usefully invoke key Marxist concepts to analyse women's position in cities and regions. Below, I give a brief summary of Marxism and socialist feminism, before examining the work of socialist feminist geographers.

Marxism and the Production–Reproduction Relation

Karl Marx and Friedrich Engels aimed to look beyond the appearance and ideologies justifying nineteenth-century European industrialisation to uncover the laws impelling capitalism. In this analysis they revealed the position of those most oppressed by this system of production—the working class. Their method was dialectical historical materialism. Marx and Engels describe the reasoning.

> As individuals express their life, so they are. What they are, therefore, coincides with their production, both with what they produce and with how they produce. Hence what individuals are depends on the material conditions of their production (Marx and Engels 1976, pp. 31–2).

Within this schema, human beings create themselves through the social relations they enter into to act upon nature and produce the necessities of life. Human labour power is fundamental in the transformation of nature into useable things. The social relations of such labour, and the ways in which the products of that labour are then distributed, vary in time and place. This variation was called the mode of production. For Marx, in the transition from the feudal mode of production to the capitalist mode of production, the means to produce goods and wealth—tools, expertise, equipment—were transferred to one class: the bourgeoisie or middle class. Deprived of the ownership of the means of production, the proletariat or working class was left with only their ability to expend their labour. This class, 'free' to sell its labour to whomever wished to buy it, embodies the capacity to create value. The generation of value occurs through the expenditure of labour power to create commodities. The commodity subsequently has two value components; both values derive from the labour power within the commodity and the social relations of exchange necessary for the realisation of that value; the two values are a use value—a utility for its producers and others—and an exchange value—a value in the market place.

Value is therefore created by the expenditure of labour power on raw materials to create usable and exchangeable commodities. Under capitalism, such a process usually occurs within a factory, shop or office during a working day. This day has two components; one part creates enough value so that, when it is transformed into a wage, it is sufficient to buy the necessities of life for a family and to support the bearing and rearing of children in that household. However, in addition to this necessary labour which is directed to social reproduction, the worker labours more, in order to produce further commodities. The value produced in this part of the working day is siphoned off by the capitalist in the form of surplus value; it is surplus value which becomes profit. The levels at which the necessary and surplus levels are set are the source of conflict and the subject of negotiation: it is in the capitalists' interests to have the surplus labour component of the working day as long as possible and in the workers' to have the highest wages so that they retain much of the products of the working day. The key sites of contradiction and ongoing class conflict arise from the negotiations around the two components of the working day and the distribution of the surplus between worker and capitalist.

In addition to these two main classes, Marx's schematic representation of capitalism has a third class. This class is only tenuously linked to the production process at any one time; and consists of the unemployed, the under-employed and the immigrant worker. These groups comprise various parts of a reserve army of labour. They act to dampen wage demands and can be mobilised into the workforce when production needs dictate.

Women, as a specific group, enter Marx's relatively abstract analysis at two points: once as a particular component of the reserve army of labour, and, secondly, as essential to social reproduction. It is with this latter set of relationships that socialist feminists have expended significant energy in their efforts to position women within Marxism. In this they were assisted by Engels' essay on *The Origin of the Family, Private Property and the State* (1975). This essay uses anthropological studies of pre-modern, pre-capitalist families to suggest that, in the eras of 'Barbarism' and 'Savagery', there were group marriages in which women had power over productive and reproductive labour as well as over descent and inheritance. At these times, there was a 'spontaneous sexual division of labour arising out of (the) physiological differences' of women as child bearers (Delmar 1979, p. 284); women controlled the domestic arena and men commanded outdoor food collection, domestic animals and the ownership of slaves. The transition from such an era to the present, class-based 'Civilization', from group to monogamous marriage, involved the 'world historic defeat of the female sex' (Engels 1975, p. 120). According to Engels, this was due to changes in the material conditions, whereby men's work—especially in regard to domestic animals—became more important than women's in providing for the needs of the household. He writes: 'Thus on the one hand, in proportion as wealth increased it made the man's position in the family more important than the woman's, and on the other hand created an impulse to exploit this strengthened position in order to overthrow, in favor of his children, the traditional order of inheritance' (Engels 1975, p. 119).

The conversion of men's greater wealth into power over women thus derived from a change in their command over resources, from a recognition of the male role in the creation of children, and an 'impulse' to control the disposal of wealth among his children. Once this power was asserted, it became oppressive to women, as women's command over domestic and reproductive labour was not only destroyed but reoriented to the service of men in a patriarchal household.

To end such oppression, especially as it was manifested in bourgeois households based on dependent women and private property, Engels suggested an extension of legal equality and the mass entry of women into public industry. Such measures were necessary preludes, he argued, to the alliance of all women with the working class to socialise the means of production, abolish private property, and usher in an age of socialist freedom, gender equality and monogamous sex love (Engels 1975).

Feminist Critiques

There have been extensive reworkings and re-evaluations of Engels's essay (see, for example, Sayers et al. 1987), and his work, along with that of Karl Marx, has been subjected to many feminist critiques. The criticisms range from the conceptual foundations of the Marxist framework to points of detail. Some feminists have seen the primary concepts of the material conditions of production and a genderless individual, as male-centred starting points. Thus Mary O'Brien suggests that there is no logical reason why social action on nature to create the means of subsistence is any more a fundamental necessity for living than the actions people take to ensure biological reproduction (O'Brien 1982). Similarly, some feminists see the apparently neutral and ungendered term 'individual' as built on a male body and conception of reason (see chapter 1). Many feminists also see the masculine assumptions inherent in Marx's abstract theorisation of capitalist society as permeating other central concepts in his work.

Thus it is possible to view the notion of class as deriving from a male experience of paid labour and as depending on the invisible presence of female reproductive labour. In the nineteenth century, many European women were not in the paid labour force; thus it was unclear theoretically, as well as pragmatically, just how Marx incorporated home-based women into his class analysis. Did they only have a class position when they were in the paid workplace? What then of the many women who only worked in the home for no wages? In Marx's schema, it seemed that such women could only share the class positions of their fathers or husbands, and that only paid labour qualified someone for a class position. As a result, women's class position was subordinated to, and subsumed by, that of men and an independent class position for a woman was attained only when they entered paid work. Domestic labour being unpaid, did not qualify as a material foundation for any particular sort of class relation, even though it was recognised, theoretically, as vital to the continuation of the capitalist mode of production.

The priority which class relations assume within the Marxist framework carries over to Engels' analysis of the family. Thus, in explaining the origin of the sexual division of labour as a spontaneous result of biological difference,

Engels has been criticised for not using historical materialism to analyse this event (Delmar 1979) and the overthrow of matriarchal inheritance. While Engels writes of a material imperative, he does not recognise or theorise the other drives which seemingly impel the change: why men apparently need to control inheritance and be assured of their paternity. For Engels, too, the sole sexual drive is to heterosexual monogamous coupling, and not the 'abominable practice of sodomy' (Engels 1975, p. 128). These drives by men for power over women, for directing inheritance, and for heterosexuality are not examined or theorised by Engels. Rather, he assumes such desires are explained by biology and does not examine their historical and material origins.

If there are theoretical problems with the Marxist account, there are also empirical ones. Thus, if the oppression of women is built upon the economic and legal power men have over them, and if that power is class-based, it follows that abolishing private property and socialising production destroys the class structure and the material foundation of women's oppression. However, the experience of socialist countries has not been encouraging in that, regardless of the legal pronouncements and forms of economic organisation, it appears that women have not yet achieved liberation in socialist states such as Cuba, Vietnam, China, states of the old Soviet Union and East Germany (Cliff 1984; Coward 1983; Einhorn 1981; Eisen 1984; Rowbotham 1972; Scott 1976). Such examples are often portrayed as deriving from misguided and erroneous applications of socialist principles; while some feminists have recently argued that the failure to liberate women in these regimes does indeed point to fundamental problems with Marxist principles.

Despite the many problems of applying the Marxist schema to an analysis of women, many socialist feminists have attempted to use the theory, as well as the substantial modifications made to address feminist concerns. This work has created a separate body of theorisation called 'socialist feminism'.

Socialist Feminism

In a widely read essay initially published in the *New Left Review*, Juliet Mitchell (1966) wrote of socialist work on 'The Woman Question': 'To this point, the liberation of women remains a normative ideal, an adjunct to socialist theory, not structurally integrated into it' (Mitchell 1971, p. 81). Mitchell's essay went on to weld Marxism to psychoanalytic theory and feminist priorities around what she saw as the four key structures of women's situation: production, reproduction, sexuality and the socialisation of children. For Mitchell, patriarchy is the symbolic law of the father, instilled from birth through socialisation processes which create deeply held and unequal gender identities. Under capitalism, this patriarchal law is primarily expressed through ideologies and can only be overthrown by a cultural revolution. If women were to be liberated from class as well as patriarchal oppression, this revolution, she argued, had to occur at the same time as the socialist revolution destroyed capitalist relations.

For the next ten years, socialist feminists focused their theoretical and political attention onto only two of the four structures isolated by Mitchell—production and social reproduction. Those working in geography shared this emphasis, and concerted work on sexuality and the construction of gendered identities had to await the reassertion of radical feminism and the entry of post-modernist thinking into the discipline (see chapters 3 and 4).

In feminist circles, efforts to include women's work into the Marxist schema went in two main directions. One attempted to theorise the place of domestic labour and the other that of women's particular place in the labour market. Thus in scores of articles, the intricacies of the labour theory of value, the nature of necessary labour and the various dimensions of social reproduction were discussed in what became known as the 'Domestic Labour Debate'. The various theoretical dimensions of the debate focused on whether housework was productive or not in the technical Marxist sense of producing surplus value; whether housework had an integral role in the capitalist mode of production or was only indirectly related to it; whether unpaid housework raised or lowered the value of the husband's labour power; whether housewives were part of the working class; whether there should be a wage for home workers; and whether housework could ever be completely socialised (Benston 1969; Fox 1986; McIntosh 1982; Seccombe 1973).

The main focus of the debate was on the possibility of Marxist categories including women (Johnson 1990a). It was not until writers like Lise Vogel (1983) and Christine Delphy (1976, 1984) centred their analysis on jobs women were doing within the home—caring for children and men rather than creating use values and labour power for capital—that women were moved to the centre of theoretical analysis. Such moves brought to the fore the issue of male power, not just class relations; they also highlighted the failure of those using Marxist categories to seriously transform their analysis in a feminist direction.

A similar shift occurred—away from Marxist categories toward feminist concerns—in that other major area of theoretical discussion: the place of women in paid work. A focus on paid work derived both from its analytical priority within Marxist theory, and from the huge changes which had occurred in the participation of women in the workforce in the West from the 1960s. This was especially the case for married women who were increasingly paid workers. It was therefore appropriate that one of the first socialist feminist attempts to theorise women in the labour force concentrated on this group. Thus Veronica Beechey observed how women's place within the family—as child bearer and rearer and as a domestic labourer paid by the male wage to reproduce labour power—both shaped their utility for capital as low cost labour and created their specific place within the reserve army of labour (Beechey 1977). Subsequent engagements with this argument have introduced sexuality as a critical element in the definition of women workers and this, as much as the material relations within the home, have come to be seen as crucial to the creation and sustaining of a sexual division of labour (see, for example, Anthias 1980; Bland et al. 1978; Cockburn 1981; Milkman 1982; Phillips and Taylor 1980; Pollert 1981; Westwood 1984).

The main focus of recent socialist feminist work has been the constitution of workplaces through a complex interconnection between home and paid labour, capitalism and patriarchy; this is also the most significant focus for socialist feminist geographers. A key debate in the USA and the United Kingdom, between 1978 and 1981, shaped the theorising of the relationship between production and social reproduction as part of a concerted attempt to link Marxism and feminism. The theoretical problem was no longer defined as fitting women into Marxist categories, but rather as integrating and trans-forming two quite separate theoretical traditions.

The result was an array of dual systems theories of two kinds: those that configured capitalism as a particular modification of patriarchal relations (such as Eisenstein 1979; McDonough and Harrison 1978), and those that saw the form of capitalist relations as essences conditioned by patriarchy (Hartmann 1981). Such dual systems are problematic: they rest on the separation of the domestic from the public, an idea which is a conservative product of capitalism. Furthermore, they tend not to account for the oppression of women as workers outside the family, as well as inside it; and they have led to a fruitless search for the historical origins of each system (Tong 1989). In addition, many empirical studies of women workers have shown the combined effect of not just class and gender relations but of race, ethnicity, place and sexuality. Identities are there-fore not divisible into single categories which somehow interact together in a composite whole, but are dynamically and multiply constituted. This view moves well beyond the socialist feminist notion of the individual interacting with structural social systems. This trajectory has been followed by socialist feminist geographers as they moved from examining women in paid and unpaid work, to looking at the interconnections between production and social repro-duction and, most recently, to seeing paid work as complexly created by a range of interconnected elements. It is relevant to turn to the way in which geographers have applied and modified socialist feminism.

PRODUCTION AND REPRODUCTION IN CITIES AND LOCALITIES

The first explicit attempts in geography to link socialist with feminist theory occurred in the pages of *Antipode* in 1973 with an exchange between Pat Burnett and Irene Bruegel. Here Burnett argued that existing models of urban form were static, conservative and assumed the sexual division of labour and patriarchal nuclear families. She saw a Marxism enlivened by radical feminism (see chapter 3) as the alternative to this ignoring of the 'structural relations in society [of]...class, sex and race'. In Burnett's analysis, which owed a large debt to Juliet Mitchell, capitalism was built upon the family, an unequal gender division of labour and a male 'psychology of power' (Burnett 1973, pp. 57–60). In a comment on Burnett's paper, Irene Bruegel recentred the analysis on Marxist theory, arguing:

> [Burnett's] analysis is rooted in bourgeois individualism and is not, as she seems to think, Marxist. For she regards 'the male psychology of power' as the dominant force in society, considering implicitly that women and men constitute opposing classes. On the contrary, class is most usefully defined, not by gender, but in terms of relationships to the means of production...to understand women's position in society and the city it is necessary to analyse the institution of private property and the role of the family, from a Marxist standpoint (Bruegel 1973, p. 62).

It was some years before socialist feminist geographers took up Bruegel's challenge. When they did, they used Engels' framework to examine the interconnections between production and reproduction at two main sites: the city and the region or locality.

Thus in 1983, Linda McDowell wrote: 'Towards an understanding of the gender division of urban space', in which she argued that production and reproduction were part of a single inseparable process which varied across space and time. Setting herself apart from those feminists doing historical analyses of the family and those engaged in the Domestic Labour Debate, she suggested that women should not be the sole objects of study but rather analysis should examine the structure of social relations that constituted female oppression at various sites. To this end she considered the development of British suburbs as the outcome of a set of historically specific, contested and actively created relations between male and female paid work, housework and domestic ideologies.

McDowell demonstrates that it was a particular interconnection of these elements which produced the postwar suburbs in the 1950s, with women out of the paid labour force and living a domestic ideal; but this interconnection also led to the fundamental alteration of the suburbs in the 1960s when women became paid workers in new, decentralised, service industries. In the 1980s, there was more change when recession led to the return of these women to the home amid a reassertion of a pro-domestic ideology, a withdrawal of much state welfare and child support, and the creation of a new informal economy (McDowell 1983).

Other feminist geographers have taken up this focus on the domestic arena as one of the pivotal sites in the negotiation of gender and class relations in the city, as well as a place of increasing importance in the generation of income. They include Isa Dyck (1989) and Suzanne Mackenzie in Canada (1987), Susan Hanson and Geraldine Pratt in the USA (1995), and Nicky Gregson and Lowe in the United Kingdom (1994). The home–work relation has also been seen as fundamental in the changing nature of employment in and across regions.

From the 1960s, the mass entry of women, especially married women, into the paid labour force of Western countries was crucial to the very creation of the Women's Liberation Movement and socialist feminism. Within geography, the work of Jane Lewis, Doreen Massey, and a series of 'Locality studies' conducted in the 1980s, led to a recognition of the importance of women's employment to the general restructuring of Western space economies. According to Lewis, the entry of women into the paid workforce had been neglected in favour of a preoccupation with 'Marxist analyses of the labour process and of processes of capital accumulation, [which] have not been particularly sensitive to questions of gender. This is reflected in the tendency to limit analysis to changes taking place within the production processes at the expense of examining the ways in which...restructuring in both production and consumption spheres interact' (Lewis 1984, pp. 47–8).

Lewis detailed the mutual effect of productive and reproductive spheres around the sexual division of labour at home, in waged work, in trade unions and in the urban and regional economies of the United Kingdom. She argued that these processes interconnected with de-industrialisation, the growth in the tertiary industrial sector and a functional restructuring of firms; and that this led to new spatial and sexual divisions of labour in the 1980s. Thus she described the growth of semiskilled and unskilled mass production and assembly work alongside the increasing importance of female labour and its availability when firms made decisions about where to locate their factories. The result was a new regional differentiation of productive activity and variable female participation rates between core areas and new green field sites, such as those in the North East of the United Kingdom where branch plants were located in the 1980s (Lewis 1984).

In a more extended analysis of this process, Doreen Massey argued too, that gender relations were important to the new space economy of Britain. Using a number of industries, including engineering and electronics, she examined changes in the nature of production, especially in its ownership, scale, internal organisation and division across space. She also documented changes in work-forces, revealing the association of men with centralised managerial, skilled and technical jobs and women with unskilled, peripheral jobs. In addition, Massey incorporated a historical, hierarchical and layered perspective, suggesting that any locality was the outcome of decisions which occurred at a global, national, regional or local scale. As a result, in Massey's analysis, any one place emerged from a series of investment decisions over time, laying down different strata of landscapes; each of which expressed various connections to global, national and regional economies (Massey 1984).

Working together, Doreen Massey and Linda McDowell examined how capi-talism and patriarchy articulated to create a set of distinctive conditions for the maintenance of male dominance at various locations across Britain. They argued that in north east coal mining villages, in Lancaster cotton towns, in the rag trade in central London, and in rural life on the Fens, capitalism presented patriarchy with different challenges at particular places; the outcome of each was a variable set of gender and class relations (McDowell and Massey 1984; see also Mackenzie and Rose 1983).

These studies outlined the way in which workforces were constituted at particular locations as a result of past investment decisions, global–local connections, work cultures and gender relations; and a number of locality studies continued this work. Using huge research funds from the British Economic and Social Research Council, geographers examined seven localities across the United Kingdom in order to uncover the ways in which specific labour markets, households and local histories came together in the process of restructuring (Anderson et al. 1983; Baggueley et al. 1990; Cooke 1989, 1990).

A study of Lancaster in the south east of the United Kingdom examined the centrality of gender relations in shaping the labour relations and character of the locality (Lancaster Regionalism Group 1985). The researchers included Linda Murgatroyd, Mike Savage, John Urry and Sylvia Walby. Thus, Savage shows how, from 1890 to 1940, capitalists mobilised patriarchal power in working-class villages to create compliant feminised workforces in Preston's cotton industry (Savage 1985). Using more recent examples, Murgatroyd describes how restructuring was conflict-ridden but also built upon the segregation of women in particular jobs in Lancaster. As a result, struggles around the sexual division of labour were seen as crucial in determining which jobs were created—whether women's or men's—and at what level of skill, status and pay. Thus for Murgatroyd, unequal gender relations shaped the capital–labour relation (Murgatroyd 1985). Theorising such relations, Sylvia Walby's words resonate with Juliet Mitchell's in 1971:

> The essential skeleton of my analysis of gender relations in contemporary western societies is constituted by a system of patriarchal relations in articulation with a system of capitalist relations...The actual pattern of gender inequality should be seen as the outcome of these two systems together with that of racism...The key sets of patriarchal relations are to be found in domestic work, paid work, the state and male violence and sexuality (Walby 1985, p. 50, see also 1986).

The earlier dual systems frameworks underlay these locality studies. These frameworks theorised the connection between capitalism and patriarchy while the locality studies added spatial, temporal and social complexity to the analysis. While many of the locality studies failed to honour the claim that gender relations would be central to them (see Bowlby 1986 and Rose 1989), they did bring to the fore the importance of production and reproduction relations in the understanding and analysis of place. However, locality studies replicated the general problems noted earlier in socialist feminist theorising. Thus, despite claims that gender and class were interconnected, analyses tended to rely on the concepts in dual systems theory; they assigned primary importance to production or reproduction, capitalism or patriarchy. Research on localities and spatial divisions of labour, such as that by Philip Cooke and Doreen Massey, tended to see class relations as primary but shaped by gender and patriarchy. Studies of particular places, such as those by McDowell and Massey, and Mackenzie and Rose, argued that capitalist class relations were built upon the more fundamental and historically prior patriarchal relations. Thus the analyses of socialist feminist geographers which attempt to connect class and gender relations seem locked into dualistic thinking.

Such end points emerge in part from the personal politics of the researchers involved, from whether they put their socialism or their feminism first. Campioni and Gross (1983) argue further that, for socialist feminists, Marxist theory requires an analysis which gives precedence to either capitalism or patriarchy. They suggest that the primacy a Marxist framework assigns to class relations and male social relations is inseparable from the perspective, and that there is no way of modifying its phallocentrism. They conclude that the socialist feminist project is doomed to replicate the social relations it purports to analyse and subvert.

Further, despite gestures acknowledging that race, ethnicity and sexuality are present in the relations of production and reproduction, socialist feminists have tended to focus on class and patriarchy.

More recently, geographers such as Katherine Gibson and Julie Graham have taken the Marxist starting point as the basis on which to build an alternate theoretical framework. This begins with the actions of people on their environment and the social relations for distributing the products of their labour. This allows the conceptualisation of a range of social relations and actions (and not just those of class and patriarchy), as well as an array of possible points for political intervention. Whether such a position remains socialist feminist is debatable. For in abandoning the focus on production and social reproduction, in moving class relations as created solely from paid labour in the public arena from the centre of analysis, this alternate framework disregards some fundamental tenets in socialist feminism. However, the new framework offers a progressive and radical viewpoint: it focuses on material relations, especially those derived from the expenditure of labour and its appropriation by others, and links this to more recent discussions of multiple identities, sexualities and the role of discourse in constituting gender (see Gibson and Graham 1992; Gibson-Graham 1996). This may be the future of socialist feminist geography.

Despite these conundrums, the socialist feminist perspective still provides a useful framework for studies of place. Thus, in a recent study of Worcester, Massachusetts, Susan Hanson and Geraldine Pratt describe the ways in which class and gender relations vary in the town's sub-regions to produce a variegated social landscape (Hanson and Pratt 1995). In this and in other locality studies, the key elements of the socialist feminist approach to geography remain. These are:

- a theoretical connection between class and patriarchy
- a recognition of the ways in which production and reproduction relations interact to produce particular places
- an acknowledgment that each place emerges from a series of investment decisions, and a set of gendered and patriarchal social relations which have occurred over time to produce a number of co-existing landscapes
- each place is also nested within a hierarchy of places (the local, the national, and the global), which also shapes the character of this place over time.

This approach informed my analysis of one place and industry in Victoria: the textile industry of Geelong.

THE PATRIARCHAL ECONOMY OF
A RESTRUCTURING LOCALITY

Following the example of socialist feminists who examined particular industries and localities in other parts of the world, this study looked firstly at national patterns of restructuring, and then at Geelong and the nature of the production–reproduction relations within its textile industry between 1984 and 1991.

Australian patterns of restructuring

As I have noted elsewhere (Johnson 1991), contemporary restructuring involves purposeful change in the ways in which capital and labour are allocated across sectors of an economy and in new engagements with the global economy. In Australia, the most striking indicators of this process have been increased unemployment, the globalisation of finance, and sectoral shifts in economic activity away from manufacturing towards service industries.

Industry	1955	1965	1975	1985	1995	1997–98
Primary –	*18.1*	*12.3*	*9.8*	*10.5*	*7.9*	*6.2*
Rural	15.9	10.4	5.2	5.0	3.3	5.2
Mining	2.3	1.9	4.6	5.5	4.6	1.0
Secondary –	*38.2*	*40.6*	*33.5*	*31.2*	*26.7*	*21.2*
Manufacturing	28.0	28.4	21.4	19.0	15.8	13.3
Construction	7.8	8.7	9.2	7.9	7.4	7.1
Exchange –	*23.4*	*23.0*	*25.7*	*26.4*	*27.2*	*27.0*
Wholesale and Retail	15.7	14.3	18.3	14.5	14.8	20.6
Transport, Storage, Communication	7.7	8.8	7.4	10.0	9.8	6.4
Public Administration and Defence				4.9	4.2	4.0
Services –	*20.3*	*24.1*	*31.0*	*26.9*	*33.8*	*34.1*
Finance & Business			15.9	11.4	16.9	14.3
Comm'y & Personal			11.2	15.9	17.5	17.4
Tourism & Recreation			15.1	15.6	16.9	2.4*

*Figures for tourism not available

Table 2.1   Employment by industry percentages, Australia 1955–98
Source: Adapted from Fagan and Webber 1999, p. 82; and *Year Book Australia* 1999

As table 2.1 indicates, there has been a marked change since 1955 in the industries in which people have been employed: this follows a decline in the rural and manufacturing sectors and an expansion in service industries. There are two main implications of such patterns: social and spatial. The sexual division of labour in Australia means that there are particular gender dimensions to these trends; these, in turn, vary by industry sub-group, occupation, hours worked, location and ethnicity. Table 2.2 offers some details of where women and men work, and how these patterns changed between 1981 and 1996–97.

	1981				1996–97			
	Males		Females		Males		Females	
	000s	%	000s	%	000s	%	000s	%
Agriculture, forestry, fishing	321.2	7.9	101.4	4.4	296.9	6.3	130.1	3.6
Mining	83.4	2.1	7.7	0.3	75.9	1.6	10.8	0.3
Manufacturing	954.0	23.3	317.3	13.7	831.1	17.5	298.8	8.3
Construction	436.9	10.7	52.1	2.3	506.8	10.7	80.1	2.2
Wholesale and retail	715.6	17.6	550.2	23.8	951.2	20.0	779.1	21.6
Transport and storage	7.2	51.7	2.2	305.2	6.4	91.1	2.5	
Finance, property and business services	306.7	7.5	252.5	11	595.2	12.5	549.1	15.2
Health, education and community services	384.7	9.5	620.6	26.8	375.6	7.9	977.9	27.0
Recreational, personal and other services	171.8	4.2	222.9	5.8	258.9	5.4	251.1	7.0
Other industries*	405.2	10	133.8	5.8	550.1	11.5	448.5	12.4
Total	4064.2	100	2313.0	99.88	4746.9	99.8	3616.6	100.1

* For 1981 these are unspecified; in 1996–97 they were electricity, gas and water; accommodation, cafes and restaurants; communication services; government and defence.

Table 2.2   Occupations and male and female employment, Australia 1981 and 1996–97
Source: *Year Book Australia* 1982 and 1997

It is clear that manufacturing, construction, transport and rural industries remain significant employers of men, and that wholesale and retailing have become important for male employment. Retailing has declined slightly as an employer of women, though women remain a large and growing presence in other key service industries; these include the health, education and community services sector and recreational and personal services. Table 2.2 confirms the ongoing sex segregation of the Australian economy.

Fagan and Webber (1999) note that the proportion of men in employment has fallen overall from the 1950s to the 1980s; net job losses since the early 1970s initially affected women more than men, but since the early 1980s, men have lost jobs faster than women. For women, manufacturing provided 19% of jobs in 1971 but only 10% in 1986; the comparable proportions for men were 25% and 18%. On the other side of the restructuring ledger, women's jobs in non-manufacturing industries have grown at a faster rate than men's, and women predominate in key service industries such as community, personal and recreational services.

Alongside these gender relations is ethnicity: thus the decline in manufacturing has affected mainly workers born overseas (especially those born in eastern and southern Europe). In the industrial workforce, workers born overseas as a proportion of employees declined from 34% in 1971 to 21% in 1986. So too, the growth of managerial and professional occupations has benefited mainly English-speaking groups; while the decline in plant and machine operators has reduced employment opportunities for workers born in Vietnam and southern Europe. Growth has been in service sectors and in the part-time labour markets, both developments which have advantaged women workers (Fagan and Webber 1999).

Restructuring is primarily an economic process (though one with profound social consequences), which is manifested mainly within the public world of paid work. However, there have also been suggestions of other revolutions in work within Australia over the last thirty years. In particular, following the logic of the socialist feminist position, it is necessary to consider the ways in which unpaid work in the domestic arena has changed. The popular media would have us believe that men are now sharing more of the housework and more of the child care than ever before. In light of changing domestic technology, the rise of married women's paid work, the decline in family size and the fall in male labour force participation, along with a rhetorical shift in the attitude of men and women to housework, one would expect there to have been a marked change in the sexual division of domestic labour from women to men. Yet the most recent research disproves this; in essence, exploding the myth. A compari-son of the Australian Bureau of Statistics' 1987, 1992 and 1997 Time Use Surveys, shows that women and men have some separate responsibilities: women are responsible for indoor housework such as cooking, laundry, cleaning and the physical care of children, while men are responsible for outdoor tasks such as the lawn, garden, pool, and pet care, and for maintaining the home and car. Shopping and playing with children are shared activities, though women still tend to spend more time in these tasks than men (Bittman 1995). A compari-son of sex equality ratios between women and men over time indicates the degree to which women remain the primary home workers, and the marginal changes which have occurred in this pattern from 1987 to 1997.

Activity	Sex Equality Ratios		
	1987	1992	1997
Laundry, ironing, clothes care	9.0	7.5	6.6
Other housework	4.2	4.1	4.1
Child care	3.6	3.6	2.8
Food & drink preparation and clean up	3.1	2.9	2.6
Purchasing goods and services	1.5	1.6	1.5
Gardening, lawn and pool care	0.5	0.5	0.8
Home maintenance and car care	0.1	0.3	0.2

* The **sex equality ratio** is derived by dividing the mean time women spent in a particular activity by the mean time men spent in that activity. If there was an equal sharing of unpaid work, the sex equality ratio would be 1. More than 1 means that women do a disproportionate share. Less than 1 means that men do a disproportionate share.

Table 2.3   Sex equality ratios in unpaid work tasks, Australia 1987, 1992 and 1997
Source: Bittman 1995, p. 9; How Australians Use Their Time 1997

Thus, in 1987, women did 9 times more ironing than men, and, in 1992, 7.5 times and in 1997, 6.6 times. In 1987, women did slightly more shopping than men, but in 1997, men did the same amount of shopping, despite the hype from a proliferation of men's magazines which exhorted them to consume more; in 1997, however, men continued to dominate in car and home maintenance. In suggesting explanations for why women in 1992 spent slightly less time on laundering and cooking, Michael Bittman notes that women are less willing to do these tasks, and that this reduction in time derives from women's attitudes rather than from any significant rise in the work men do in the laundry and kitchen. Similarly, when the labour force and household work is disaggregated, it is clear that there are major differences in the average amount of time males and females spent on these activities. Males spent 18.7% of their day on labour force activities and 10.4% on household activities, compared with females who spent 8.8% on labour force and 20.2% on household activities. Despite the major shift in the core economy then, there have been few changes in the domestic economy or, perhaps, in the relationship between the two. To explore this idea further, it is useful to assess how these two economies affect different localities.

Figure 2.1a Shopping man
Source: the Age, *Saturday Extra*, 12 October 1991

Figure 2.1b Domestic woman
Source: Mary Leunig 1982

Restructuring has occurred unevenly in Australia. Melbourne and Geelong in Victoria, Wollongong and Newcastle in New South Wales, and Whyalla and Port Pirie in South Australia were once important sites for manufacturing. These states and cities have been deeply affected by the decline in manufacturing. On the other hand, new regions and cities have either come into existence or expanded as a consequence of various developments. These have included resource developments in Perth and parts of Western Australia, the expansion of tourism on the Gold and Sunshine Coasts in Queensland, and the growth of business services in Sydney, New South Wales. The rest of this chapter will examine one of the centres affected by the decline in manufacturing industry, namely Geelong in Victoria, in order to detail the restructuring of paid and unpaid work in one locality.

Restructuring Geelong

Restructuring in Geelong both typifies the manufacturing sector as a whole and also reveals the specific ways in which its historical layers of investment, and the social relations which derive from them, have characterised this locality. The socialist feminist framework for this study (Johnson 1990b and 1991) focuses on the production–reproduction relation, the sexual division of labour and a particular notion of a patriarchal economy; by this I mean that set of social relations revolving around the expenditure of labour and the appropriation of its surplus which systematically advantages men over women. This concept links production to social reproduction and patriarchal relations, and sees these relations as an interlocking system. Examining the recent restructuring of Geelong, the fate of particular industrial sectors and the constitution of one workforce over time, also means taking account of other factors. These include connecting the global to the local, and an appreciation of different rounds of investment and how they come together to create a place-based patriarchal economy.

Geelong is 80 kilometres south west of Melbourne. It is Victoria's second largest city, and was initially established in the 1830s as a port for exporting wool and grain. It became an industrial centre in the 1860s with locally owned woollen textile mills, hide processing plants, and soap, rope and candle manufacturing. Multinational capital entered in 1925 with the arrival of the Ford Motor Company from the USA. Ford was to expand its operations in the 1940s and 1950s, and was joined by other overseas companies involved in the car, truck (International Harvester 1939), aluminium (ALCOA 1960) and petrochemical industries (Shell 1964). These companies were lured to this provincial city by offers of free land, quality road and rail infrastructure, payroll tax exemptions, state provided housing, and an available and skilled workforce, which was both Australian and overseas born. In addition, local but regional companies established textile, clothing and footwear plants in a city which had become known as the 'Bradford of the South' (*The Melbourne Star*, 1935), a centre for skilled textile workers, training and research. By 1970, following these investments, Geelong was very much a manufacturing city with its industry large scale, primarily masculine and overseas owned.

Entering Geelong for the first time in 1979, the nature of these investments and activities were all too evident, but so too, were some of the changes which had occurred. I passed public housing and large manufacturing enterprises—the Ford Motor Company's six-cylinder engine plant, the Shell Oil Refinery and a range of component suppliers for the car industry, such as Henderson Springs and Pilkington Glass. Around the scenic bay front were stockpiles of fertiliser, wheat silos, wharves and, across the waters, the towering smoke stack of the aluminium smelter. In addition to this productive landscape, there were empty buildings, ghosts from earlier phases of investment, such as the wool stores on the bay front and, along the Barwon River, a number of textile mills. Since 1979, a newer landscape has overlain this older one, as Deakin University has expanded, government offices have decentralised from Melbourne, research institutes have been established (in education, textiles, marine science and animal health) and tourist–retail attractions have been constructed, such as the Bay City Plaza, the National Wool Museum, and Steampacket Place on the waterfront (see figure 2.2).

In the 1960s and 1970s, the above-mentioned industries employed quite different workforces. A relatively rigid sexual division of labour in this conservative regional centre has ensured much lower rates of female workforce participation rates than the Australian average. Most of the women who obtained paid work entered the textile, clothing and footwear parts of the manufacturing sector. In contrast, the very large industries, such as the car and truck plants, the construction industry and petrochemicals, employed Australian-born and migrant men.

In their changing pattern of employment loss and closures, these industries echo the experience of Australia as a whole. Employment losses and closures have affected the sexually divided workforce in different ways, as figures 2.3 and 2.4 show. These figures also show the other side of restructuring. Alongside the massive contractions in male manufacturing, construction and utility employment since the 1970s, male employment has increased notably in the service sector: especially in the 'Wholesale and Retailing', 'Finance and Business', 'Communication', 'Community Services', 'Public Administration', and 'Entertainment and Recreation' sectors. Far more significant, however, has been the growth in female employment in the service sector, consequent on the doubling of the workforce participation rate of women, the growth of part-time work in this highly feminised sector as well as the relative decline of the textiles, clothing and footwear industries.

Why have these patterns occurred? We need to connect this locality with the relevant global networks of capital investment and disinvestment, to engage

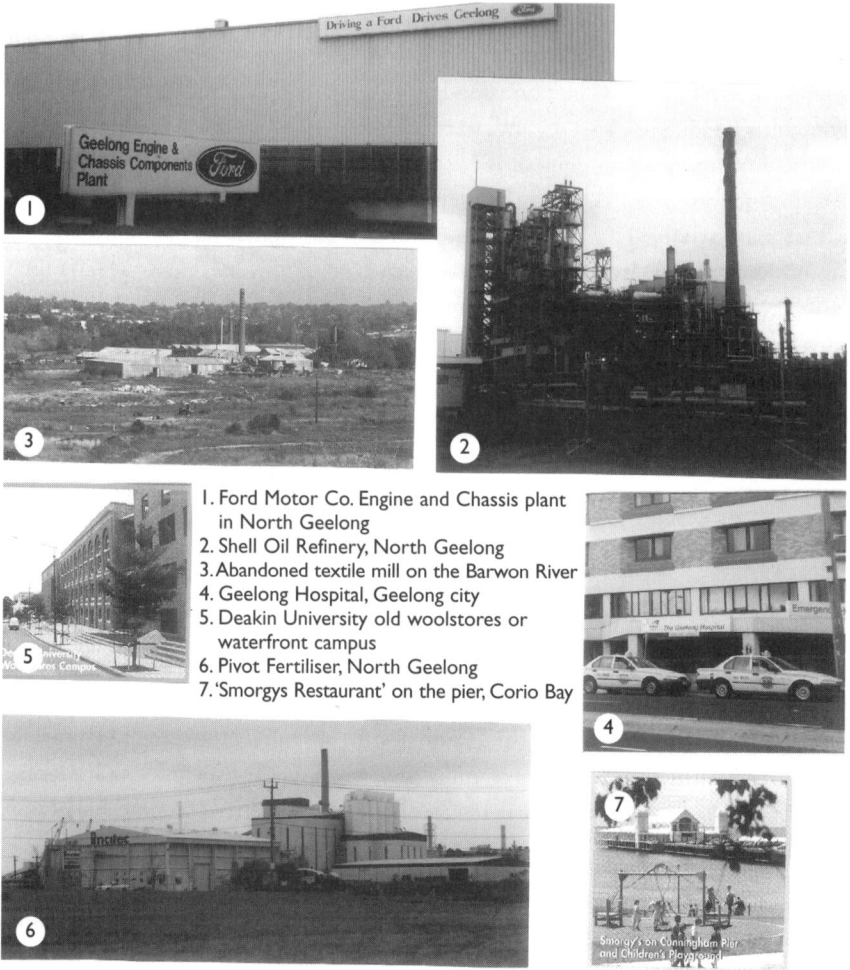

1. Ford Motor Co. Engine and Chassis plant in North Geelong
2. Shell Oil Refinery, North Geelong
3. Abandoned textile mill on the Barwon River
4. Geelong Hospital, Geelong city
5. Deakin University old woolstores or waterfront campus
6. Pivot Fertiliser, North Geelong
7. 'Smorgys Restaurant' on the pier, Corio Bay

Figure 2.2 Geelong's industrial, service and tourist landscape
Source: Geelong Regional Commission and Louise Johnson

with the general economic and spatial trends for Australia as a whole, and to acknowledge the role of various levels of government activity in directing economic change. We also need to understand how local decisions, social relations, institutions, histories and geographies create a restructuring experience which is particular to this place.[1]

The graph in figure 2.5 shows the phases of job displacement as occurring in four moments of restructuring. In Geelong, each moment was associated with a particular phase of heightened job loss: in 1974 to 1975, these losses were primarily in textiles, clothing and footwear; in the early 1980s, they occurred

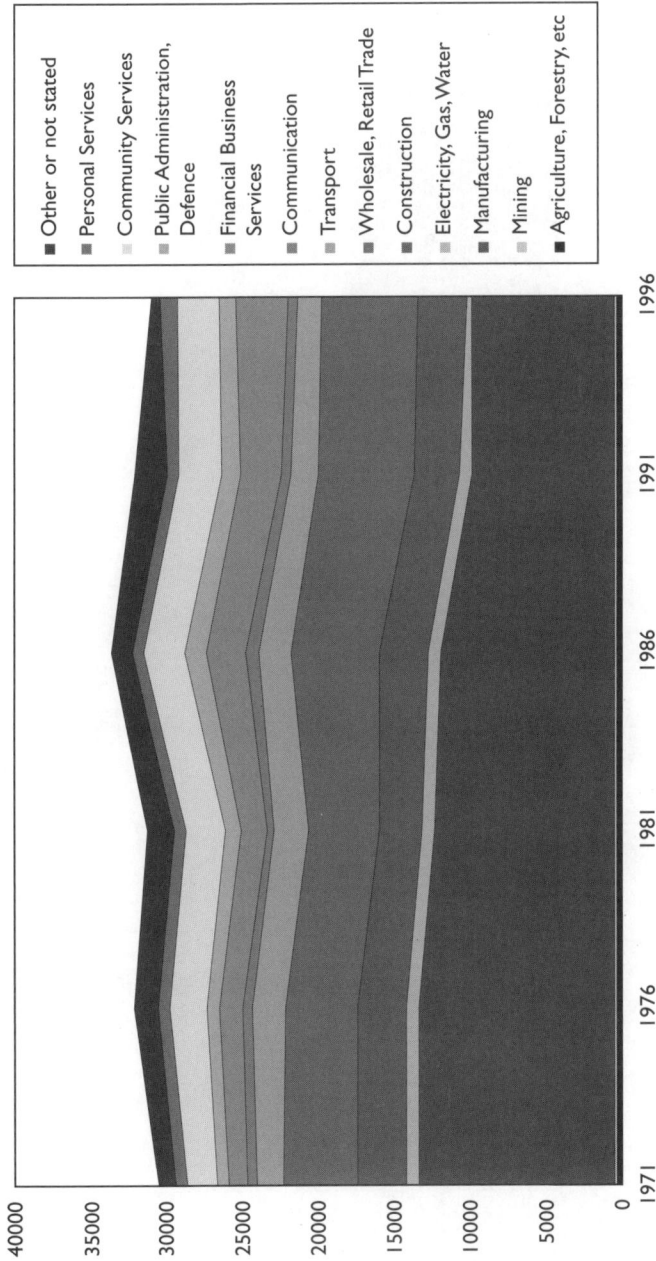

Figure 2.3 Male employment, Geelong 1971–96
Source: ABS Household Censuses 1971, 1976, 1981, 1986, 1996 Census of Population and Housing catalogue no. 2020.0 Time Series Community Profile.
Graph by Cathie Newton

Industry—Female Employment Geelong Industry 1971–1996
Source: Australian Bureau of Statistics Census of Population and Housing

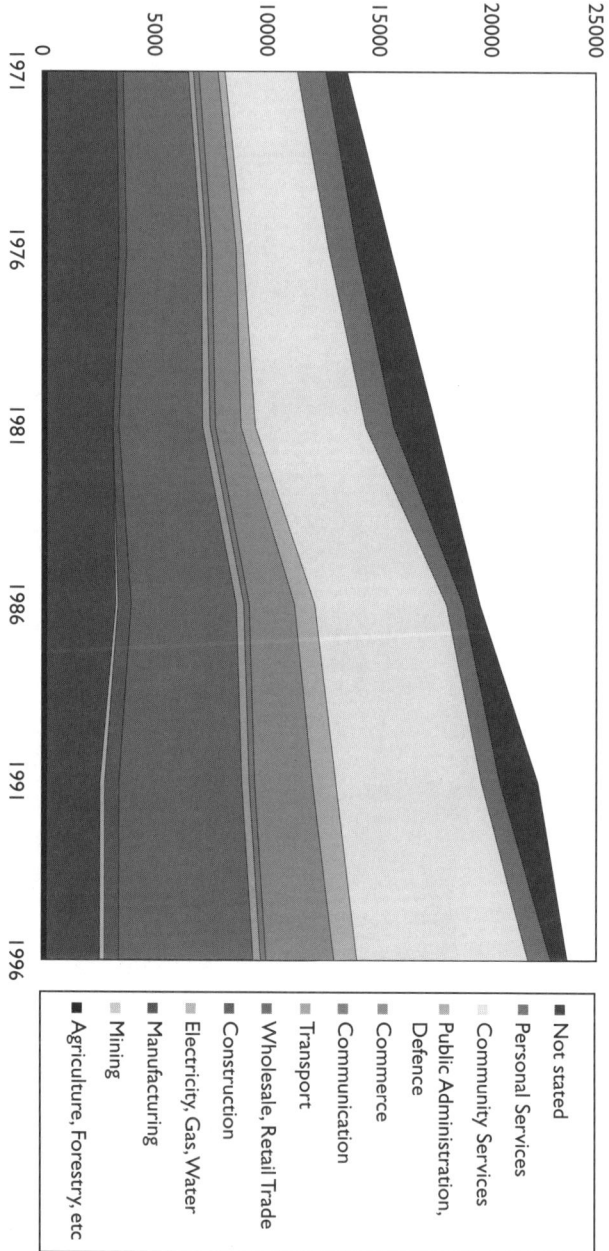

Legend:
- Not stated
- Personal Services
- Community Services
- Public Administration, Defence
- Commerce
- Communication
- Transport
- Wholesale, Retail Trade
- Construction
- Electricity, Gas, Water
- Manufacturing
- Mining
- Agriculture, Forestry, etc

Figure 2.4 Female employment, Geelong 1971–96
Source: ABS Household Censuses 1971, 1976, 1981, 1986; 1996 Census of Population and Housing catalogue no. 2020.0 Time Series Community Profile. Graph by Cathie Newton

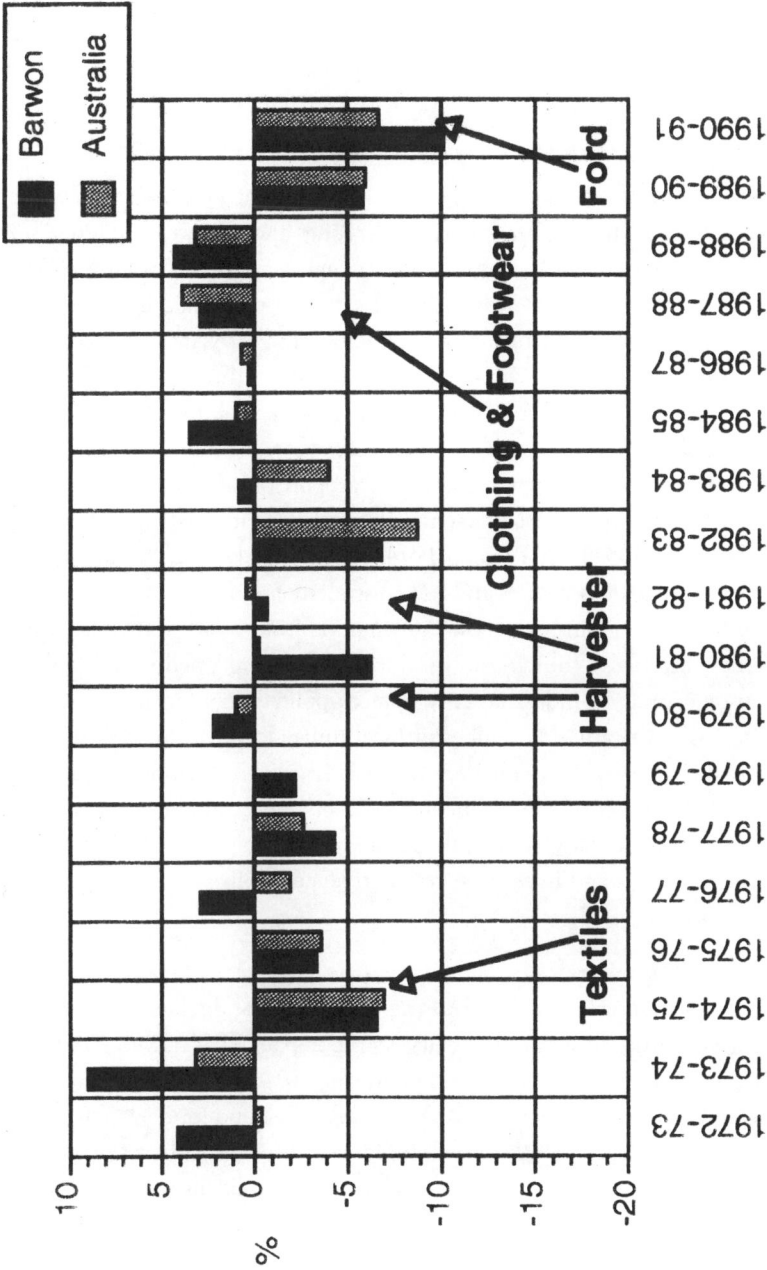

Figure 2.5 Employment change indicating four moments of economic restructuring, Geelong 1972–91
Source: ABS Censuses 1971, 1976, 1986, 1991, 1996. Graph by Steve Wright

when the truck and agricultural machinery maker International Harvester closed down; from 1989 to 1991, when Ford downsized its Geelong plant, and textiles again contracted; and, most recently, when the public sector was cut. Each of these moments had specific social and spatial impacts.

1. Textiles, clothing and footwear, 1971–74

From 1972 to 1974, a number of Federal Government policy shifts occurred which drastically affected the textile, clothing and footwear industries in Geelong. The Equal Pay decision of 1972 (which was phased in from 1972 to 1975) finally recognised that women had long been underpaid in these industries and elsewhere. In instituting this principle, the Arbitration Commission directed that over the following two years there be three pay increases for women. This directive was crucial in changing the sexual division of labour in the local textile industry: over the next ten years, the proportion of men employed in these industries increased; the proportion of women employed in these industries fell from 60% of the textile workforce in 1961 to 40% in 1986 (Johnson 1990b, 1991). In 1973 to 1974, there was also a 25% cut in tariffs across a wide range of industries and the Australian dollar was devalued by 17%. The result was a huge influx of low cost imports from South-East Asia, where the textile—especially the clothing—industry had burgeoned under multinational auspices and pro-industrial development policies. For Geelong, the effect was swift and devastating: a number of local mills closed down and others cut their workforces. Thus between 1971 and 1974, the 3000-strong textile workforce was halved. Most of the affected workers were middle aged, married, Anglo-Celtic and migrant women, the latter, mostly from southern Europe, had arrived in the 1950s and lived close to the riverside mills in Chilwell, east and south Geelong.

2. International Harvester, 1980–83

The second restructuring occurred in the early 1980s and fell most heavily on the workforce of the International Harvester Company (IHC). Since 1939, the plant had been a major employer, and by the late 1970s it had 2500 employees. In 1980, IHC laid off 390 workers. Between June 1981 and June 1983, when the firm's local tractor plant finally closed, it reduced its staff numbers from 1100 to less than 50 (GRC 1981, p. 2). In part, Harvester's problems were those of the domestic agricultural machinery market: falling demand, increased overseas import competition and the removal of a Federal government bounty (GRC 1981, pp. 30–1). But a more important factor leading to these job losses, were the extreme difficulties the parent company was experiencing in Chicago. Characterised by a company historian as a 'marginally profitable, overly diversified, and thinly capitalised corporate monolith', Harvester had nonetheless registered a record business year in 1979 (Marsh 1985). However, within a matter of months, its management had triggered a long and bitter strike over compulsory overtime and seniority in its US factories: this was immediately followed by a worldwide lessening in demand for heavy agricultural equipment. In 1980, sales fell 24%, and in the next two years, the company lost more than $US2 billion (Jury 1983).

At the same time, the parent company's Australian subsidiaries were making a number of poor decisions. Local management, imagining the boom in resources meant a rosy future for the company, borrowed heavily to lift production. While US managers were cutting spending, selling assets and preparing to renegotiate its $3.4 million debt, stockpiles in Geelong continued to grow to the point where no more credit was extended either in the USA or in Australia. By 1983, the multinational had closed more than a third of the plants it had operated worldwide four years earlier, and in the process had cut its staff from 98 000 to 32 000 (Marsh 1985, p. 278). In this corporate downsizing, the Geelong plant was a minor casualty, but the consequences for its 2000 strong, mainly overseas born and male blue collar workforce, were devastating. Many of IHC's former workforce found employment with the car maker Ford and associated industries—the manufacturing sector still being able to absorb experienced skilled workers—until this company too, began to retrench workers in large numbers and, by the early 1990s, it was taking on few new employees.

Accompanying this tide of retrenchments in the early 1980s was another wave of female displacement from the textile, clothing and footwear sectors, as the Federal Government hastened the pace of tariff reform.

3. Ford Australia, 1990–91

Under pressure from the so-called 'Button Plan' to rationalise its operations, and with a declining share of the domestic car market and a growing stockpile, Ford stopped production for a number of weeks in late 1990. In December, management cut 210 white collar positions at its Geelong plant, then, in January 1991, 850 production jobs, mainly in the skilled trades area (*Geelong Advertiser* 1991). A number of Ford's local suppliers began their own lay-offs; the 1991 Census recorded a decline since 1986 of more than 1600 positions in the local 'Transport Equipment' sub-sector, a net loss of 31%. Ford has since secured a deal to manufacture engine blocs for export to Mazda plants in Japan, but continuing automation and the failure of new models to gain more market share has made any return to employment levels of the 1970s unlikely.

Information on who was retrenched from this plant became publicly available during negotiations surrounding the honouring of superannuation entitlements. Newspaper and union accounts suggest that most of those who lost their jobs were middle aged, overseas born, skilled men. Ford had recruited many of these men in Italy, Malta and Croatia in the 1950s. In the early 1990s few anticipated working again. A significant number of Australian-born men were retrenched, but those with many years' service in the plant received severance payouts; these payouts allowed them to pay off houses and to establish businesses. However, further details on these workers—obtained via local networks and an unemployment centre—suggested that their plans to set up in horticulture, and mixed businesses or to retrain for employment in the aerospace industry had foundered, and that most of these men, in their late forties and early fifties, had entered involuntary retirement.

Both Ford and International Harvester were located in north Geelong, and many of their workers were accommodated in purpose-built public housing units. A number also lived in other working-class parts of Geelong, such as in South Geelong, West Geelong and Grovedale; a few lived on the Bellarine Peninsula. Most, then, came from those parts of the city which had long been characterised by low income and education levels; from areas where there were higher levels of unemployed and of workers with trade qualifications. Geelong therefore conforms to restructuring and social polarisation patterns noted for Sydney, Melbourne and Adelaide, in that the areas most affected are those with already low incomes, high unemployment and poor educational levels (see Baum and Hassan 1993; Fagan 1986; Johnson 1996a, chapter 1).

4. Services, 1993–95

The fourth episode of intensified restructuring, between 1993 and 1995, was in the previously expanding tertiary sector. With Federal, State and local governments set on reducing expenditure and workforces, there have been major job losses in health, education and community services. Many of these dismissals have been disguised as 'voluntary redundancies', so that detailing these retrenchments is difficult: some of the employees now provide services in this sector on a subcontractual basis, and some have been re-employed. However, it is still possible to isolate a total of 1200 retrenchments, in the years 1993 to 1995, in education, the port and water authorities, and the hospitals, and in Federal, State and local government bureaucracies and planning agencies.[2]

This final moment of restructuring has affected women more than men, the unskilled as well as the para-professional, and parts of the city, such as south of the Barwon River, which have not previously experienced high levels of unemployment. Thus the effects of this and other restructurings have variable class, gender and spatial dimensions.

This overview of restructuring in Geelong has confirmed the importance of looking at general trends as they are manifested in any one place. It has also shown that:

- it is important to situate a locality in the global fortunes of capital, especially in a city which has large manufacturing plants owned by multinational corporations
- the various arms of the State have direct impacts on a locality: through setting up a national trade, currency and industrial regulatory environment, and in providing and withdrawing services and employment
- a local built and social environment is the outcome of successive waves of decisions to invest or disinvest
- in this environment gender, class, ethnicity and space are each important in creating a workforce. These elements combine in particular ways to shape the character of an industry—be it textiles, car making or education—the jobs to be done in those industries, and the experience of retrenchment. Thus the sexual division of labour is central to the creation of workforces, industries and jobs.

The rest of this chapter looks at the way in which the patriarchal economy operates at a local level. Relations of production and reproduction can be seen by looking at the creation of one workforce, and the ways these relations are deployed in one manufacturing plant.

Creation of a Textile Workforce

The textile industry has a long and distinguished history in Geelong. In the 1860s the town boasted the first wool processing operation in the newly independent colony of Victoria. The first mill was established here because of a classic combination of natural resources, capital, labour, government encouragement and expertise. There were abundant natural resources—the Western District fine merino wool clip, soft water for processing and waste disposal—and British equipment and expertise. There were also local capitalists seeking to employ the post-gold rush population, and a government bounty had been created to encourage urban industry and to absorb the potentially revolutionary fervour of the post-Eureka population (Cope 1981; Linge 1979). The wool industry so created was hampered by poor management, under capitalisation, and flooding, as well as competition from quality imports from Britain. Nevertheless, it slowly expanded until the First World War: thereafter, apart from the 1930s, it grew at a greater rate, with larger mills, a diversification of fibres away from wool, new producers—such as carpet makers—a growing concentration of ownership, and, given high tariff protection, greater output (Wheelwright and Miskelly 1967; Worrall 1987). The boom conditions lasted until the contraction of the Geelong industry in the early 1970s.

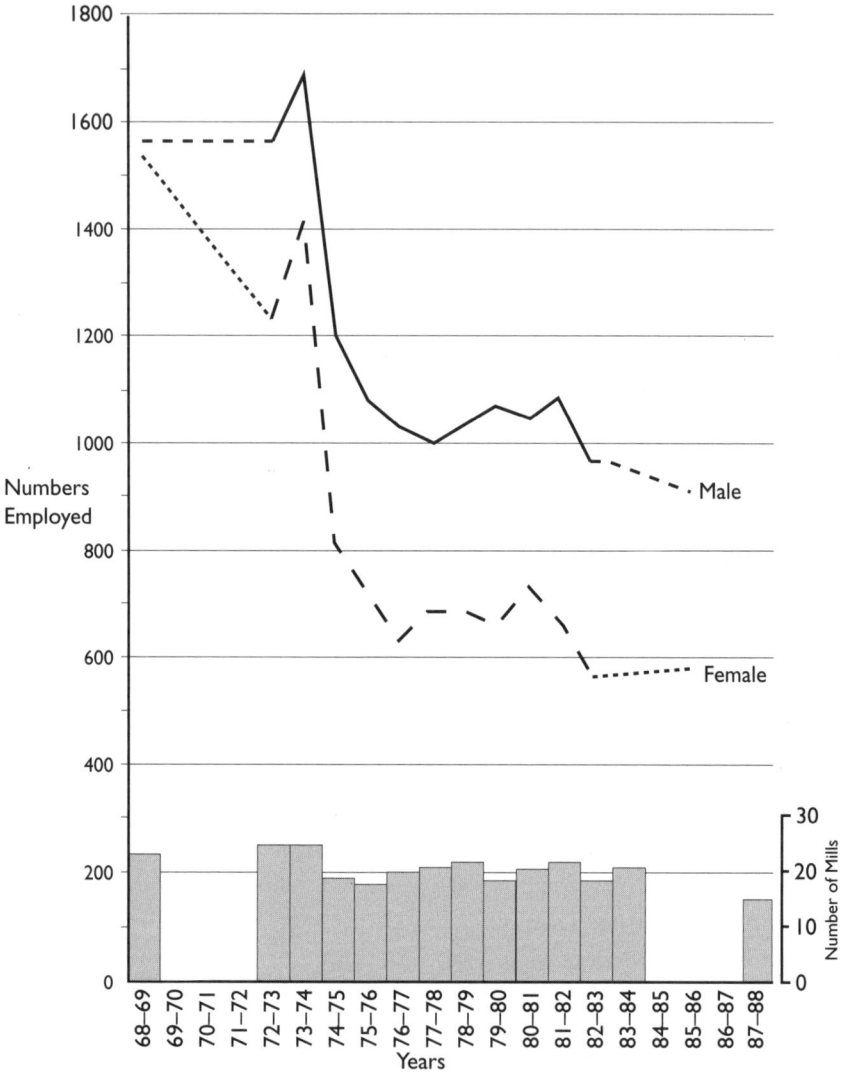

Figure 2.6 Textile plants and employment by sex, Geelong 1968–88
Source: Johnson (1991)

During the 1960s, two carpet manufacturers arrived in the city, adding to the existing seven wool-based textilers; one of the new arrivals was a knitter and yarn maker, and the other a specialist felt producer. There were also numerous takeovers, a range of technological developments as well as mounting dilemmas for the mainstay of the city, which was woollen and worsted production—as it confronted mounting international competition from rival fibres and producers.

The 1973 federal government decision to relax trade restrictions, escalated this encounter, but it was the equal pay decision which shaped the form of the response. Figure 2.6 shows that, as the number of mills and the textile workforce declined, so too did the proportion of women in the industry. There are a number of interconnected reasons for this change: the equal pay decision meant women were no longer cheaper employees; secondly, those parts of the textile industry which expanded, namely the manufacture of carpets and the production of specialist felt, mainly employed men, while those parts which had mostly employed women—namely the production of woollens and worsteds—contracted; thirdly, there was a technological and ideological shift between 1970 and 1988. Comments from employers in 1988 reveal some of the reasoning. For example:

> Yes. Equal pay affected us. It suits us better to have men in some sectors, especially in carding and spinning. You can't expect women to work hard and do the lifting. Otherwise you have to get expensive machines.

> We have a sex ratio of 50:50. We are employing fewer women, especially in weaving. With the new looms you need fewer operatives and more technical people. The type of work needs lots of supervision (in this mill and elsewhere by male technicians and foremen).

In this mill and in others, technological changes thus led to fewer female operatives, and more male overseers and technicians. This changed the nature of the work, and who was doing it as well as the relation of women and men to it. What is being described here is a shift in the sexual division of labour between home and work—so that more men and fewer women are being employed in the industry—and within the plants themselves, as men are associated with jobs which are either expanding in number or they are taking over 'new' jobs arising from technological changes. In contrast, women were seen as a 'problem': they were now too expensive, could not lift heavy weights and were unfamiliar with the technology. In a survey of Geelong's textile mills in the late 1980s, it was apparent that men tended to be associated with managerial, overseeing, design, technical, sales and labouring work. With the exception of the latter, these types of jobs increased in number and proportion during the 1980s. Women were in areas where they had worked since the 1960s, namely winding, spinning, weaving and mending. All these tasks were undergoing technical changes or were being reduced in importance in the 1980s.

The case of one mill, surveyed in 1982 and again in 1988, gives further insight into how the production–reproduction relation was constructed and negotiated in one plant as it was restructuring. Valley Worsted was established in 1924 in

South Geelong by two drapers from Adelaide and one from Melbourne. In the 1960s and 70s, it underwent a series of takeovers, ultimately becoming part of a large, Sydney-based investment company in the early 1990s. From a workforce of over 600 in 1960, 70% of whom were women, there were 170 workers in 1982, 59% of them women; and in 1988, there were 120 workers, of whom only 45% were women. Where these women worked and the ways in which their work–home relations were negotiated was quite different from the men in the plant. These connections affected who got work, the status of that work and its associations with technology, skill and pay. The patriarchal economy of this plant was observed and described in interviews with 100 workers in 1982 and 1988[3] (only a few of whom had been interviewed in 1982).

These interviews revealed that, in 1982 and 1988, the meaning of the separation between home and work varied for women and men in apparently identical class positions. These meanings shaped decisions on whether or not to enter the labour force, the ease with which jobs were obtained, the nature of the experience of both paid and unpaid work, work histories, and what work was done where, when and for what remuneration.

Home–work relations
A dye-house operator typifies the attitude most men in the mill held in 1982: 'I like to work because I should work. My wife, she doesn't work.'

This man felt the profound responsibility of supporting his family and home. In contrast, younger, unmarried men were less concerned with working; they said with pride, that they worked to get cash for cars and pleasure; once they married, the need to work, at whatever job, became crucial. This need was never questioned: it was necessary to support the family, to buy the house or to 'keep the wife'. The working class was something, then, that men should and did occupy. For women, however, it was a transgression, as the following comments confirm:

> This is the first time she's worked since we were married. I don't like it. I like her to be at home. We don't need the money (Male, weaver, 1982).

> My husband was horrified at me working after I first got married. He believed I should be home to look after the kids...He was driving a truck at Hirsts. He was asked if he knew any menders to work 9 to 3. He told them that, even though his wife was a mender, she wouldn't want to work. That started me going! Now he doesn't have a choice about me working (Female, mender, 1988).

Despite these male anxieties and resentments, most men admitted that the woman's wage was essential for the household, though women's paid work remained something to be explained and justified in a way never deemed necessary of men's wages. Male management and workers saw women workers as interlopers and treated them accordingly.

All the married women 'had to work', either because their husbands were unemployed (an increasing problem in this city where manufacturing was declining) or their wages were low; similarly, widows and women separated from their husbands had to work. As one woman noted: 'I really didn't want to go back to work after my last child but with my husband unemployed, I had to work. But then I just kept going. I enjoy my job' (Warper 1982). This illustrates the nature of these married women's involvement in the paid workforce: it was both a positive choice and a necessity. It is also primarily for the family: 'You work for your family. You try to be everything for them first. You always put them before house and work' (Female, mender, 1982). Most men, however, saw their prime responsibility as that of breadwinner bringing money into the household, with few responsibilities beyond that. Most men perceived home as a haven where women worked and they relaxed, even if their partner was in the workforce. Their contributions to cooking, cleaning, washing, child care or shopping were usually described as 'helping' and as given grudgingly when asked. As one mender put it: 'I do all the housework. My husband doesn't cook, though he helps dry the dishes. My son is too busy and doesn't want to...I get tired.' (Female, mender 1982). The nature of this working-class life differed for the men and women; for men received domestic support and women gave it.

Family connections were most important in obtaining and leaving a job. Many women learnt of a job through a family contact, or a friend who worked for the company or 'just knew' a job was coming up. These contacts help explain the ethnic make-up of this plant. Many of the workers had been at Foster Valley for many years; they came mainly from the adjacent Anglo-Celtic working-class suburb of South Geelong. The creation of this workforce, the 'contacts' through families and friends, and management's preference for hiring people known to existing employees, ensured the continuity of this workforce. Quite often, two or three family members, and sometimes entire families, worked in the mill. Thus the mill had a very local character.

When women left their jobs, this was most often due to their family orientation: they might leave because of their child-care responsibilities, to look after an aged or sick family member or to follow their husband's work move. Men usually left 'for work reasons', 'choice' and 'ambition'. Management shared these views of the home–work relation, and this affected their hiring and dismissal practices. Thus in 1988, management replaced female with male weavers, portraying this move as a 'logical' outcome of new equipment; it was also, however, an exercise in removing 'troublesome' and 'unreliable' single and married women from the plant in favour of the truer and needy occupants of the jobs, namely male breadwinners. Married women were also encouraged to take extended holidays and were put onto part-time work when times were bad. Both managers and workers saw women as being able to cope far better with losing their jobs than men. Women who lost their job could return to a legitimate—if low status—underpaid role at home where they had an identity and meaning. Male interviewees saw life without paid work as nothing: there was no place to go which offered solace, status or legitimate activity.

The findings from these interviews on the relation between home and work can be summarised as follows:

Men:
- have to work
- their wage is primary
- their domestic responsibilities end with the pay packet, the garden and the car
- have no alternative to work: they actualise themselves in the public sphere
- occupy classes; in the 1980s, they belong in paid work.

Women:
- both choose to, and have to work
- their wages are vital but secondary
- their domestic work includes child care, cooking, cleaning, washing and the second income
- their alternative to paid work is home: a private sphere which is equally rewarding
- enter the 1980s mill working class as interlopers.

The gendered workplace

As I have observed of mills in Australia and the United Kingdom, there was a relatively rigid sexual division of labour in this mill. While it may not lead to identical patterns of work across mills and is a pattern which is both contested and unstable, in this mill the division between male and female work was regulated by words, spatial layouts, dress codes and equipment.

Men:
- dominated management. In plush, quiet offices, they held most power, wore suits and were on the highest salaries
- predominated in middle management, in production, in the office, in sales and design. In these jobs they wore suits, and white or grey coats, depending on whether they were in the office, on the shop floor or in a laboratory. Their work was designated as skilled and they were often formally educated; in 1982, these men had qualifications in textiles, while in 1988, more had qualifications in accounting, management and computing

- in the dye house and wet finishing areas, in contrast to the white collar areas, wore overalls and work boots and prided themselves on the hard, dirty and dangerous nature of their work where they had to lift great weights and deal with heat and steam
- also dominated the stores and dispatch where they were engaged in and defined by the heavy lifting involved.

Figure 2.7 Men's textile work, 1984–88
Source: Louise Johnson

Women:
- performed 'routine' administrative reception, sales and clerical work. Poorly paid, these white collar workers were still regarded as 'staff', dressed smartly and shared the separate office spaces with non-shop floor middle managers
- on the production side, were concentrated in two areas: machine minding and mending; in 1982, machine minding involved combing and winding, warping

and weaving, and in 1988, only weaving. The weaving and winding rooms were noisy and mechanical; here men did the 'technical' work and women the 'unskilled' machine minding. The mending room was a quiet all-female space, where women did 'unskilled', close sewing work on large elevated tables arranged in neat rows. In both places women wore aprons. Women also worked in the stores and dispatch area by 1988.

Figure 2.8 Women's textile work, 1984–88
Source: Louise Johnson

Socialist feminists have been particularly concerned with the way in which jobs are categorised as women's or men's, and have highlighted the complex, dynamic and contested nature of the process. In analysing the sexual division of labour in this Geelong textile plant from 1982 to 1988, five key elements emerged:

1 The attribution that men, but not women, possessed skill and specialist knowledge and were familiar with technology

2 Men who were supervisors and managers assumed authority and status over some men and all women; all men assumed an air of authority over women

3 The definition and regulation of gendered bodies in space. Thus management saw men's bodies as strong and physical, a perception the men took care to reinforce; while management saw women's bodies as characterised by innate, natural characteristics which included nimbleness, intolerance of heat and an inability to lift weights

4 These perceptions were reinforced by the associations of particular tasks with these notions of masculinity and femininity. Thus men's work was manly, necessary, far removed from domestic labour, well paid, mobile, interesting and essential; women's work was clean, safe, routine, sedentary, and not unlike housework in being boring and repetitive, a perception reinforced because these women wore aprons

5 A sense of the inevitability but also the questioning of these associations.

The existing patterns of employment were rarely if ever questioned. The clear sexual division of labour in the mill was usually seen as 'the way things were' and always would be. Challenges from the trade union were unheard of; its concern was saving the industry and the jobs in it; and, resistance to the work regime from smaller collectivities of workers mainly took the form of unseating an unresponsive shop steward or dealing with health and safety issues. However, in small ways, individual women did challenge the way things were: by naming their inequitable position in relation to housework, by speaking against their lack of high level training and the low value assigned to their work. The women also talked of the pleasure as well as the pressures of doing their work, of their pride in their high (if unrecognised) levels of skill, and of knowing when management was wrong and standards were falling; they also enjoyed the sexual banter with men which enlivened their days. This knowledge and these actions gave them some power in situations where they formally had no power, and allowed them to challenge some of the foundations of the particular sexual division of labour in the plant. However, these small gestures were weak in the face of the larger forces of a patriarchal economy which were buffeting this mill.

Over the six years this plant was studied the main change was the increase in male labour. The fall in the number of women employed can be directly related to the entire removal of some centres of female work such as the spinning and winding sections. The two areas of expansion, those of finishing and weaving, are more interesting.

Finishing was seen as a process which fitted well in the new company structure, one which allowed commission work to boost internal production needs and the full use of new and expensive equipment. However, this equipment and the processes involved in wet finishing were clearly associated with hot, wet, heavy and technical work. There was therefore no question that women would be employed to replace masculine dominance in this domain. Finishing was an area where more men were being hired, where its association with male labour was being affirmed and where longer hours were being worked.

Since the invention of the power loom in the nineteenth century, women have been cast as weavers. Weaving requires reliability, nimbleness and a supposedly 'natural' inclination to work yarn into cloth; these qualities have long been associated with women's bodies and temperaments, so that women have been seen as uniquely suited to weaving. However, in 1988, when a conservative management gave men the new jobs in this area, weaving shifted from female labour to male. This redefined and mobilised a new array of gendered associations. Men's breadwinner status received renewed emphasis, and the new looms were portrayed as more technical and needing greater skill and physical strength to operate, despite their being smaller and lighter than the ones they replaced; however, these looms used synthetic rather than natural fibre and men were seen as better able to work these 'heavier' synthetic fibres than women. The need to run these machines twenty-four hours per day, also meant that only men were considered for them; management saw women's moral well-being and the sanctity of their role at home as making them unsuitable for night shift work. Thus management saw men rather than women as having the moral option, the bodily and mental capacities and the need to work these new machines. Women were recast as unreliable, unskilled and morally unfit for the new weaving machines, as well as not needing the work.

In this way, management, in renegotiating a variety of existing patterns and gendered assumptions, in deciding who worked and where, whose work was skilled and had what level of status and authority, affected the sexual division of labour both at home and work The interconnection of class and patriarchy, home and work, production and reproduction was fundamental in the restructuring of this plant.

This extended discussion of one workplace in a Victorian locality demonstrates the efficacy of the socialist feminist approach in geography. But two things remain problematic.

First, socialist feminism has attempted to recognise relations other than those of class and gender which structure home and work, but these relations tend to be neglected and are not fundamental to its theoretical or empirical project.

Thus my own account, along with many other socialist feminist geographies, has inadequately dealt with the nature of identity, which includes ethnicity, race and sexuality; these aspects can affect the creation of a workforce as fundamentally as class and gender. Chapters 4 and 5 take up this point.

Second, the notion of a patriarchal economy is relatively untheorised and tends to put male power above that of class power in a dualistic system. The interconnection between the two elements—patriarchy and class—results from the articulation of separate entities. Class has been extensively theorised by socialist feminists, but patriarchy has been admitted without extensive consideration. Derived from radical feminism, it is a concept which has been debated widely in feminist circles and in geography. The next chapter considers this critical concept in the development of feminist geography.

Notes

1   This quest began in 1984 and, until 1990, took the form of research into the textile industry for my doctoral dissertation. Since 1992, my research into this place has been funded by Deakin University Arts Faculty, the Australian Research Council, and the Australian Housing and Urban Research Institute. It has encompassed a number of projects: 'Post-Fordist Geelong', 'Restructuring the Locality', 'Social Polarization in Geelong', and 'Servicing the Service Sector'. It has also involved a number of co-researchers, and the funding support of these various organisations is gratefully acknowledged.

2   Jason Brandrup secured figures on people displaced from government agencies through a number of letters, phone calls and personal approaches to staff in those agencies.

3   In researching my PhD, I first approached all ten Geelong textile mills in 1982 for access to their workforces. Only one agreed to this request: Foster Valley. In 1982 I spent most of four months in their mill, and talked to 63 people. In 1988, I conducted 40 interviews over four days. These represented 35% of a sample, and consisted mainly of those willing to speak. In both 1982 and 1988, all sections of the mill were covered.

3

Radical Feminist Geography

In the early 1970s political positions in the Women's Liberation Movement began to diverge. Some feminists saw themselves as working within socialist theory, some as in the liberal tradition, while others called themselves radical feminists. Radical feminist groups, like the Redstockings of New York City, were responsible for the first 'policy statements' of this new self-conscious group within Women's Liberation. The 1969 Redstocking Manifesto declared that:

> Women are an oppressed class. Our oppression is total, affecting every facet of our lives. We are exploited as sex objects, breeders, domestic servants, and cheap labour. We are considered inferior beings, whose only purpose is to enhance men's lives...Because we have lived so intimately with our oppressors, in isolation from each other, we have been kept from seeing our personal suffering as a political condition...We identify the agents of our oppression as men...Men have controlled all political, economic and cultural institutions and backed up this control with physical force...All men receive economic, sexual and psychological benefits from male supremacy. All men have oppressed women...We regard our personal experience, and our feelings about that experience, as the basis for an analysis of our common situation...We identify with all women...We repudiate all economic, racial, educational or status privileges that divide us from other women...We call on all our sisters to unite with us in struggle (Tanner 1970, pp. 109–11).

Shulamith Firestone, who was a founding member of the Redstockings, spelt out this position in one of the first book-length feminist texts; there she provided an extended theoretical analysis of the origins, nature and future of women's oppression. Firestone's *Dialectic of Sex* (1979, first published in 1971), emphasising biological reproduction, and Kate Millett's *Sexual Politics* (1977, first published in 1969), focusing on patriarchy and male violence against women, became key texts in the development of an identifiable radical feminist position.

While the early Women's Liberation Movement was impelled more by radical feminist ideas than by liberal or socialist ones, the impact of this framework on feminist geography has been slight. Indeed the debate within geography of radical feminism's core concept—patriarchy—has centred on whether the term could be redefined in terms of Realist philosophy and successfully incorporated into socialist feminist geography. The one clear example where radical feminist writing and politics has been critical to geographical study has been work on the safety of women in the city. This 'city of fear' literature highlights both the strengths and limitations of the radical feminist approach with its focus on women as a united and oppressed group, patriarchal culture and male violence. In this chapter, I will outline aspects of radical feminism before considering the two main ways in which it has entered feminist geography; these are the debate on patriarchy and the city of fear literature.

Radical feminism has been historically vital to the contemporary women's movement and the analysis of patriarchy, male violence, biological reproduction and child rearing; in addition, it has brought to the fore the issue of sexuality, especially the power relations which surround normative heterosexuality and the definition of 'deviant' sexualities. The third part of this chapter extends this outline of the radical feminist position in feminist geography by examining some of the ways in which heterosexuality and homosexuality have been inscribed onto the Australian urban environment.

RADICAL FEMINISM

Reproductive Politics

Kate Millett's *Sexual Politics* formalised a number of ideas then current in the Women's Liberation Movement. She argued that, in all known societies, the relationship between the sexes was based on male power over women, a power which overrode class and racial divides. This patriarchal power was instilled in childhood and affirmed in the education system, in the family, and in the misogynist literature of writers such as D.H. Lawrence, Henry Miller, Jean

Genet and Norman Mailer. Patriarchy, 'the institution whereby that half of the populace which is female is controlled by that half which is male' (Millett 1977, p. 25), is also enforced by economic exploitation and the use and threat of force. For Millett, these patriarchal social relations persisted over time and space, though precise manifestations varied (Millett 1969).

In contrast to Millett, Shulamith Firestone begins her analysis of women's position with Marx and Engels. However, she then inverts their analysis to focus on gender relations and those surrounding biological reproduction. Her book draws parallels and connections between patriarchy and science, rationalism and technology. She traces patriarchal dominance back to the role that men assumed as protectors of nursing mothers. She argues that historically, pregnancy and child dependency placed women at a disadvantage to men and made it possible for men to wield power over lactating mothers and subsequently all women. If, as Firestone argued, the fundamental social division on which all other forms of exploitation are erected is women's reproductive capacity, then it follows that as this element is altered—through different forms of parenting or reproductive technology—then the material basis of male dominance would vanish. Foreshadowing later technical developments and their radical feminist critics, such as Corea (1985) Rowland (1992–3), Firestone (1979) also noted the remarkable persistence of patriarchy across history, suggesting that any real change in the basic male–female relation of dominance and submission could not be guaranteed to follow radical changes in reproductive technology.

Extending Firestone's analysis of the connections between biological reproduction, male anxieties about paternity and patriarchy, Mary O'Brien—a former Glasgow midwife—focused her analysis on the acts of copulation and childbirth. She argued that for men, the experience of copulation is an alienating one: that it separates them from nature, their own seed and the certainty of paternity. By contrast, women are certain that the child they bear is their own, and by bringing this child into the world by active labour, women maintain a connection with nature and their child: the child is thus a product of her unalienated labour. Men, however, respond to their alienation and uncertainty by oppressing women through a number of institutions; these include the family, the law, the church and 'morality', all of which confirm their paternity and power over women (O'Brien 1981, 1982).

The politics surrounding child bearing, and especially child rearing, have been the subject of extended radical feminist analysis; this has focused on the ways in which mothering reproduces the patriarchal order and on how new reproductive technologies have extended rather than ameliorated male power. So, for example, Adrienne Rich distinguishes between the biological fact of bearing children—which she glorified—and the patriarchal regulation of mothering through laws, conventions and the medical profession. It is the latter, she argues, which combine to create pain, distress and alienated women. If such institutions could be destroyed, Rich (1976) suggests women could think and create a utopian future through their bodies and the experience of birth. While Rich examined the effect of mothering on women, Dorothy Dinnerstein (1976) and Nancy Chodorow (1977) drew on psychoanalysis and psychology to look at the ways in which motherhood affects society as a whole. They argued that the female monopoly of child care was at the heart of women's present problems; proposing greater male parenting as the way to forging new egalitarian gender identities so that women could enter the public sphere as equal participants (see also Eisenstein 1984).

But if Rich, Dinnerstein and Chodorow see in mothering and male parenting the possibility of women's liberation, a number of Australian feminists have followed local developments in reproductive technology with growing anxiety. Thus Renate Klein (1989) and Robyn Rowland (1992–3) have documented the ways in which women's bodies have been used for experimental purposes and, they argue, colonised by male doctors in the name of 'curing' infertility and 'giving women what they want' in the form of a healthy baby. Rowland argues that women's bodies have been treated as Living Laboratories, subjected to dangerous exploratory techniques, induced to produce multiple foetuses and undergone massively invasive and dangerous procedures. She asserts that these techniques and the medical research into flushing, cloning and in vitro fertilisation which has accompanied them, have meant that women have involuntarily participated in a male fantasy of reproductive control; and that this is underpinned by misogyny and womb envy (Rowland 1992–3).

These developments in reproductive technology and their radical feminist critiques, are a long way from the creative and liberatory possibilities which Shulamith Firestone envisioned. They tend to confirm two key elements of the radical feminist position: the dogged persistence of patriarchy and male violence against women. It is these two dimensions, rather than issues surrounding biological reproduction, which feminist geographers have taken up.

The Politics of Male Violence

Throughout the 1970s and 80s, radical feminist studies documented the horror and violence associated with men's oppression of women. In these two decades, feminists detailed the misogynist character, and the incidence of pornography, rape, the traffic in women and the definition of madness. Thus, in 1972, psychologist Phyllis Chesler examined the statistics on American mental patients and their clinicians. She found that the predominantly male profession of psychology confined many more women than men in mental institutions. Her study of these 'mad' women (like that of Jill Matthews (1984) in Australia) concluded:

> Women are impaled on the cross of self-sacrifice. Unlike men, they are categorically denied the experience of cultural supremacy, humanity, and renewal based on their sexual identity...In different ways, some women are driven mad by this fact. Such madness is essentially an intense experience of female biological, sexual and cultural castration, and a doomed search for potency. The search often involves 'delusions' or displays of physical aggression, grandeur, sexuality and emotionality...Such traits in women are feared and punished in patriarchal mental asylums (Chesler 1972, p. 31).

A few years later, Susan Brownmiller described, in some detail, the historical and geographical incidence of rape. Her conclusion rested not only on the massive incidence of rape but on the culture of fear and humiliation it created for women. She states: 'From prehistoric times to the present, I believe, rape has played a critical function. It is nothing more or less than a conscious process of intimidation by which all men keep all women in a state of fear' (Brownmiller 1976, p. 15).

Kathleen Barry (1979, 1995) sees this culture of fear, violence and humiliation as underpinning an international traffic in women for the purpose of prostitution, slavery and pornography. Similarly, Susan Griffin examines the ways in which pornography constructs women's bodies as mastered, bound, silenced, beaten and even murdered by the 'chauvinist mind'. Griffin contends that this mind has been separated from nature, whereas women maintain a link with nature, peace and their selves. She argues that the pornographic mind hates and fears this link; the pornographer strives for control over nature and hence over women: it is this mind and striving, which dominates our culture (Griffin 1978, 1979, 1981). Thus for radical feminists, pornography joins other forms of violence against women—such as reproductive technology, sexual slavery, 'mental illness', war and rape—to ensure female subjugation.

Sisterhood and Separatism

While radical feminists see men and patriarchy as the enemy, their strategies for changing that dominance vary. Some political strategies involve changing the power and technical relations present around birthing and mothering; others emerge from the anger and pain surrounding male violence and require women to act collectively to fight for rape law reform, to 'reclaim the night' or to change the definition of mental illness. All of these movements emerge from pain and anger and also from a valuing of women's experiences. With a respect borne from the ethics surrounding consciousness-raising groups—which involved hearing, validating, and sharing women's experiences—went an unbounded faith in the capacity of women to link together, to heal each other, and to form effective political alliances. In these actions, women were seen as united across class, race and ethnicity by their common gender oppression to form a sisterhood. For radical feminists, this sisterhood must be separate from men and male culture, the separation being essential to the creation of an autonomous women's movement and, where necessary, to a separatist politics.

Separatism ranges from a complete severance of all sexual and social relations with men to strategic separations in order to build up a sense of female power, solidarity and alternative culture. In formulating such a notion, some radical feminists argue that heterosexuality is both a fundamental institutional founda-tion for patriarchy and a major barrier to the creation of women-centred communities. Thus in 1980 Adrienne Rich wrote 'On compulsory heterosexu-ality and lesbian existence' in order to affirm the critical role played by hetero-sexuality in the constitution of the family and male control, to expose the neglect and violent suppression of lesbianism, and to argue for a continuum of women-centred relationships; these were to range from passionate friendships (an idea Janice Raymond (1991) develops) to lesbian separatism. Rich describes the ways in which institutions such as the church, family, schools and the media assume and therefore normalise sexual relations between men and women. She also outlines the ways in which such institutions act to police heterosexuality as the norm, vilifying alternatives and giving men unlimited and unquestioned sexual access to women. Rich questions the range and extent of such practices in terms of both what is being suppressed—relationships between women—and for whose benefit. She argues that these practices affirm male sexual, economic and political power over women, and simultaneously delegitimise relationships between women. In contrast, Rich asserts the positive value of female-centred relationships and points to their contemporary, historical and cross cultural presence (Rich 1980). She writes:

> The assumption that 'most women are innately heterosexual' stands as a theoretical and political stumbling block for many women. It remains a tenable assumption, partly because lesbian existence has either been written out of history or catalogued under disease, because it has been treated as exceptional rather than as intrinsic, and because it is an immense step to

acknowledge that for many women heterosexuality may not be a 'prefer-ence' but rather something that had to be imposed, managed, organized, propagandized, and maintained by force (1980, p. 648).

What follows from this analysis is a questioning of heterosexuality as norma-tive, and an examination of the various institutions and spatial arrangements which reinforce 'compulsory heterosexuality' for both women and men. This is one direction that a radical feminist geography could take.

Rich joins other radical feminists in suggesting that while many women's lives have been adversely affected by patriarchy, they have also found spaces in which to resist and develop both their own selves and an essential female culture. A strategy of separatism leading to the recovery and celebration of a woman-centred culture underpinned Mary Daly's *Gyn/Ecology* (1978). This text is an exercise in writing women's culture, of taking words and recasting them, of reclaiming lost meanings—for terms like hags, witches and spinsters—and of developing an alternative women's language. It is also a detailed exposé of the misogynist horrors Daly sees as characterising patriarchal culture across time and space. Among the practices she examines are Chinese foot binding, Indian widow burning, African genital mutilation and American gynaecological prac-tice: each instance a way in which men's power and desires force women to mutilate and even to kill themselves. From such studies Daly posits a female-centred world, to be created by those women who share her knowledge and her vision. It is a radical feminist vision of a woman-centred utopia, populated by those who have rejected patriarchal culture, male violence and heterosexuality.

The emphasis in Daly's work on male violence, compulsory heterosexuality, a women's culture and patriarchy are elements which remain in contemporary radical feminist scholarship (see Bell and Klein 1996). There are, therefore, a number of core beliefs which distinguish radical feminism in both its 1970s and 1990s manifestations:

- women are a social group united in their oppression by men
- patriarchy—as the structure of male oppression—is the primary problem for women
- individual men as well as men as a social group benefit from patriarchy
- radical feminism is created by and for women
- radical feminism is woman-centred
- the personal is political
- 'theory' derives from women's experience; it informs and emerges from personal and collective reflection and liberatory praxis
- women need to and have created an autonomous women's liberation move-ment, a progressive sisterhood

- analysis and action are focused on those sites where male power and violence over women is most in evidence—in reproduction, in sexual assault, pornography and compulsory heterosexuality
- the objective is a non-patriarchal world, a utopia where gender difference does not matter and where women are separate, woman-focused and powerful (Rowland and Klein 1990, 1996).

Within such broad radical feminist principles, there are a number of different emphases. In her analysis of patriarchal societies, Shulamith Firestone argued that women were defined and exploited by men primarily because of their reproductive capacity (Firestone 1979). Among those who explored this concern with the role of women as mothers, bearing and rearing children in ways which reproduced patriarchal social relations are Adrienne Rich (1976), Nancy Chodorow (1978), Mary O'Brien (1981, 1982), Dorothy Dinnerstein (1976), Renate Klein (1989) and Robyn Rowland (1992–3). Other radical feminists have focused more on providing evidence, and analysing and alleviating male violence (Chesler 1972; Dworkin 1976; Daly 1978), especially in the form of rape and sexual assault (Brownmiller 1975; Griffin 1978, 1979), pornography (Dworkin 1981; Griffin 1981) and sexual slavery (Barry 1979). Additional focuses have included issues relating to sexuality, especially heterosexuality as both socially constructed and compulsory (Rich 1980), and an affirmation of woman–woman relationships, such as passionate friendships (Raymond 1986) and lesbian sexualities.

Given this theoretical and historical agenda, radical feminism has been deemed to have a number of serious limitations. Socialist feminists claim that the emphasis on women as a group united solely by male oppression ignores (or at least downplays) the importance of class and material relations for women (see chapter 2). Feminists from black, indigenous or minority positions frequently see the focus on gender as ignoring the role of imperial and racial relations in Western societies (see chapter 5). Other feminists see the withdrawal into separatist politics and communities as de-politicising the feminist agenda (Eisenstein 1984). There have also been criticisms of the relatively essentialist assumptions of radical feminism in its emphasis on women as a group oppressed by a similarly homogenous group of men, groups distinguished primarily by their biological sex. In an era where identity is now regarded as far more fluid, contextual, performative and without any necessary connection to biological sex (see chapter 4), this viewpoint is seen as anachronistic. Contemporary radical feminists (such as Rowland and Klein 1996) dismiss such criticisms as unfounded; nevertheless, they remain real issues for those in geography who, inspired by radical feminism's rhetorical flourishes in the early

1970s, now work in a climate heavily influenced by socialist and postmodern feminisms.

Some of the earliest liberal feminist work in geography drew its inspiration from the ideologies and actions of radical feminists (see chapter 1). This included their key concerns—such as patriarchy and sexuality—which have been critical to research on women's place. However, the impact of radical feminism on feminist geography has been slight, perhaps because many female geographers came late to feminism, when much of the energy and excitement of radical feminism had dissipated; when socialist feminism dominated in Britain, and liberal feminism in the USA, both feminisms having a contradictory relationship with radical feminism; and when postmodern theorisations of sexuality, gendered identity and desire (see chapter 4) challenged the entry of the radical feminist agenda into geography. Nevertheless, in the 1980s, the key radical feminist concept of patriarchy entered Anglo-American geographical discourse, as did the issues of male violence against women and sexuality. The following section discusses these feminist geographies, and their various debts to earlier radical feminist work.

PATRIARCHY AND 'CITIES OF FEAR'

Patriarchy

In 1986 Jo Foord and Nicky Gregson published an extended essay on the meaning of 'patriarchy' in *Antipode*; the foremost international journal on Marxist and other radical positions in geography. There followed a major debate in geography as to the meaning and utility of the term. Foord and Gregson saw the conceptual focus of patriarchy as crucial to feminist geography and as owing much to radical feminism: 'we argue [that] it is patriarchy, [which is] the only feminist concept which isolates the common thread to women's subordination—men's domination—which has the most potential in the development of a feminist theoretical framework...Our specific contribution here is to undertake a reconceptualisation of patriarchy, based on realist methods of abstract analysis' (Foord and Gregson 1986, p. 187).

Foord and Gregson firstly reconstruct a history of the many meanings of patriarchy in feminist discourse. In this they primarily engage with socialist feminist definitions, usages and debates around the term, thereby ignoring extensive radical feminist discussion and placing themselves firmly in the socialist feminist tradition. They end by suggesting that the diversity of definitions of patriarchy results from and produces confusion, chaos and a lack of theoretical rigour. Such variability they designate as a problem, one which can only be solved by the clear and logical thinking offered by realist philosophy. In their quest for an all-embracing, rational, cogent, clear and workable definition, they locate themselves in a modernist theoretical tradition focused on the search for certainty, a search now regarded as both futile and untenable (see chapter 4).

In one of four replies from the many international responses published in response to this article—an impressive number for any journal to carry on one article—I questioned this characterisation of patriarchy. I asked whether we should question their quest and see it as part of the patriarchal knowledge industry: 'Their objective to create an overarching theory to explain all women's oppression using criteria and a concept somehow detached from history and social context puts their discussion into the realm of patriarchal knowledge, not in opposition to it' (Johnson 1987, p. 212). The problem, as I then saw it, was not only the search for the meta-theory but also their choice of vehicles, that of realism.

In their 'Comments on Critics', Foord and Gregson clarify what they took from realist philosophy: a conceptualisation of the world as inter-related but with analytically discrete structures; an analysis of causation which suggests that objects possess causal powers but these powers are only activated in particular periods and places; the use of rational abstraction to determine the basic characteristics of an object; and the necessary and contingent relations between objects (Gregson and Foord 1987, p. 372). This analysis creates a hierarchy of abstraction: from the general to the particular and thence to the individual instance; its schema isolates necessary or basic aspects of causality, and, for any term, its contingent or historically variable elements which are also part of causality. They do so to distil what they see as vital and logically necessary, as well as variable elements in patriarchy. For Foord and Gregson, patriarchy is a particular form of gender relations. Gender relations characterise all societies and within them a set of necessary elements exists: these are male and female genders, biological reproduction and heterosexuality. Patriarchal gender relations are particular to specific societies; in the 1980s, this particularity is characterised by contingent forms of male domination which are primarily asserted within marriage and the nuclear family.

By designating species reproduction and heterosexuality as necessary to any patriarchal gender order, Foord and Gregson made them historical or universal, a claim I and others criticised. Thus I argued that, while there has to be a differentiation between women and men for any gender order to exist and biological reproduction is obviously necessary for the continuation of any human society, only these elements can be posited as necessary in a patriarchal gender order. As many lesbian mothers can attest, heterosexuality is not necessary for biological reproduction. Furthermore, the meanings of masculinity and femininity and the construction of sexuality are highly contested (Johnson 1987). Gier and Walton similarly argued that gender is culture bound. They cite hermaphroditism and androgyny as two of the more obvious cases where the distinction between male and female is unclear and problematic (Gier and Walton 1987). Lawrence Knopp and Micky Lauria also argue that heterosexuality is contingent rather than a necessary element in the patriarchal gender order, in other words, its form and meaning vary historically (Knopp and Lauria 1987).

As to how patriarchy relates to capitalism, Foord and Gregson see biological reproduction as the key element in women's subordination. Arguing from their socialist feminist position, they suggest this role articulates with class society in particular ways to create the spatial and temporal variations in gender relations, and that these are what geographers should study. Their argument is a dualist one which privileges patriarchy over capitalism; though they also argue that the two elements shape each other in historically specific ways. In contrast, Linda McDowell's comment puts a classic Marxist feminist position: she explains women's location in capitalism by a class analysis. She extends Engels' analysis to argue that women's oppression is based on the role they play as reproducers of labour power (see chapter 2), a role that primarily benefits capitalists rather than men as a group. She suggests that it is the provision of subsistence to women during the child bearing period, and not the sexual division of labour, that forms the material basis of women's subordination in class society—a subordination which is quite different for middle-class as compared with working-class women (McDowell 1986). Knopp and Lauria take issue with the priority accorded class analysis; they suggest that patriarchy has to be conceptualised not as an independent system but as a set of social relations that are continually adapted to, and articulated through, a system of class relations shaped and changed by capitalism (Knopp and Lauria 1987, p. 51).

The 'patriarchy debate' and responses to it revolved primarily around the efficacy of realist philosophy as a vehicle for conceptualising patriarchy, and how this formulation sat with various socialist feminist positions. Far from introducing radical feminist ideas to geography and placing patriarchy at the centre of feminist geography or clarifying its meaning and utility, the debate strengthened socialist feminist geography and its various research agendas around paid and unpaid work. Thus, in a contentious and limited way, the *Antipode* discussion introduced a key feminist term into geography.

Another aspect of the radical feminist agenda—the awareness of male violence which underpins patriarchy—was taken up in studies of women's anxieties in the city surrounding sexual assault and the resulting limitations on their mobility. Building on work done by radical feminists in sociology, women's studies and criminology, this 'City of fear' research has become a focal point in geography for radical feminist concerns.

'Cities of Fear'

The geographer Susan J. Smith has been instrumental in bringing a spatial perspective to the understanding of crime. In 1987, Smith introduced geographers to some of the national crime surveys undertaken in the USA and the United Kingdom, arguing for the importance of a geographical perspective— with its emphasis on neighbourhood, community and the spatial organisation of social relations—to an understanding of the fear of crime (Smith 1987, p. 1). In her review of existing research, Smith highlighted patterns in the class, location, age, race and gender of perpetrators and victims of both property and personal crime, as well as the discrepancy between actual crime rates and the fear of crime. In particular, she noted that 'while fear increases with age amongst men, it is pervasive amongst women at all times of life' (1987, p. 7). She cites studies of women's fear in cities which show how this restricts women's mobility and leads to strategies of avoidance and caution. She reports how women are forced to retreat into and fortify their homes, to avoid certain parts of the city and to curtail night time activity. She notes that 'in Britain as a whole, one fifth of women in the inner city attribute their hermit-like existence to the fear of crime...in Islington as many as half the women (yet only one sixth of the men) always or often avoid going out after dark...a figure matched in a recent survey of Southwark's housing estates' (Smith 1987, p. 13). Smith suggests a range of measures which can ease such fears: accurate and non-sensationalist media reporting of crime and greater neighbourhood and social integration: both can heighten feelings of control, safety and surveillance.

Smith's overview and geographical perspective extends the sociological and criminological literature (such as Gordon and Riger 1989; Schepple and Bart 1983; Stanko 1990; Warr 1985) on women's fear and experience of crime. Her matter-of-fact approach and non-feminist politics, however, cast her work outside a radical feminist rubric. A radical feminist view of men's violence against women and women's fear in the city has a genealogy that moves from the work of Susan Brownmiller and Susan Griffin in the late 1970s to that of Jalna Hanmer and Mary Maynard in England in the early 1980s. This genealogy differs from the generation of data on women and crime, and provides an alternative approach which recognises that the very fear which is being researched undermines the accuracy of any national crime survey derived from using door-knocking, questionnaires or phone sampling. In contrast, Hanmer and Maynard use a series of community-based meetings and qualitative research methods to document from the victims a range of male actions against women: from wolf whistling and sexual harassment through to domestic assault and rape. In their own West Yorkshire community, talking as women to women in small groups and extending the definition of 'crime' to any form of male violence which was felt or observed (rather than defined by the law as a crime), produced astonishing figures on the incidence of male violence against women. They found that 60% of the women they talked to had experienced or witnessed male violence; in particular, sexual harassment (59%), physical violence (16%) and threats (24%) (Hanmer and Maynard 1984).

In their definition of the problem and their approach to its research, Hanmer and Maynard take a radical feminist position. When analysing their data they argue that male violence is a circular system in which women are victimised, confined to their homes and placed under male protection in the public sphere. They assert that fear is fed by the media, informal rumour, personal experience and stories shared by women. As a result, women are expected to be more careful in their public behaviour than men. Women's fear of public abuse is thereby socially sanctioned and increases the belief that the home is (or should be) a safe place. This encourages women to become more home-centred, more dependent on men and participate less in public. Hanmer and Maynard argue that this is a closed, self-perpetuating cycle which can only be broken if women take control and develop ways of depending on each other for protection. They assert that such a strategy is a radical (feminist) one because it does not depend on men for its success. In their analysis, men are the enemy and only women's actions on their own behalf can deal with the problem of male violence (Hanmer and Maynard 1984).

The spatial implications of Hanmer and Maynard's work, as well as that of the criminologists who preceded them, has been variously taken up. The fact that women have geographies of crime and fear has profound effects on their travel patterns and on their experience of space more generally. Thus the 'Geography of fear' literature includes studies which describe how women tailor their travel patterns, even their employment, to feeling safe on public transport. In this context, Lynch and Atkins discuss a survey by the Women's Committee of the Greater London Council; the survey showed that only 37% of women felt safe travelling after dark by bus, 17% by the underground and 15% walking (Lynch and Atkins 1988). Because they feared using public transport, women chose to work closer to home and not to work at night. These anxieties also constrained women's other activities. In a survey of the inner Adelaide suburb of Unley Iain Hay reported that, despite the relatively low level of exposure to crime or threatening behaviour, one in five of the 1046 respondents stated that they had curbed their activities, especially at night, because they feared violence. This was the case for 32% of the women but also for 7% of the men. Eighty per cent of the women indicated that at night they feared and were unwilling to walk to the nearest shop, use a public phone, walk to a friend's house, walk though a neighbourhood park or walk home from a local cinema, bar or restaurant. Forty per cent of men expressed similar fears (Hay 1995).

Thus work by geographers and criminologists has shown that there is both a geography of crime and a geography of fear. Interestingly, it has been the dissonance between the two, and the fear itself, which has received most scrutiny from radical feminist geographers, thereby affirming Susan Brownmiller's (1976) dictum that the fear of rape, as distinct from the actual experience of rape keeps all women oppressed.

Among radical feminist geographers, Gill Valentine's work has been the most sustained engagement with this subject. Her research in Reading, England, of eighty working-class and middle-class women has detailed the ways in which their fear was constructed and manifested (1989), how such fears were then circulated and acted upon (1992), and how it could inform the better design of public space (1990). Locating her work within the criminology, sociology and feminist work on the subject, Valentine—following Hanmer and Maynard—notes how the elevation of the public sphere as a place of inherent danger for women both debars them from a legitimate place in public and gives a false sense of security in the home; in fact, as Valentine notes, home is where the vast majority of rapes and acts of male violence against women occur. How this comes about she details through her analysis of qualitative interviews, studies of gender socialisation and the media. She asserts that, by disproportionately publicising attacks committed in public places rather than domestic violence, the media connects the general dangers women fear to the public environment, and links crime with particular locations such as parks and railway stations. The media does not suggest women who are abused by their partners should not date or live with them, only that they should avoid 'dangerous places' (Valentine 1992).

Having noted the media focus on public spaces and the need to recognise the prevalence of domestic violence, Valentine joins a number of other feminist designers and planners (such as Matrix 1984; Leach et al. 1986) in suggesting measures which can improve women's sense of safety in public places, which might remove (or lessen) limits to their mobility and pleasure. She notes how women in Reading were particularly anxious in multi-storey car parks, on public transport, in open spaces and on closed-in pathways; in places that were frequently deserted, had blind spots or offered difficult escape routes. She suggests ways of reducing or relocating entrance ways, of providing more lighting in public spaces, of opening up and removing blind corners and subways, and of altering landscaping. She also argues for the clearing of derelict areas and more ground floor development (Valentine 1990). In response to this survey and more localised ones—such as that carried out in Sydney's Liverpool—a range of strategies and urban design suggestions have emerged. These have included better lighting around railway stations, designing suburbs and shopping areas to facilitate people gathering and thereby increase the passive scrutiny of open spaces, and removing blind corners, underpasses and alleyways.

In Australia a comparable study was completed by an Honours student—Melissa Permezel—who talked to groups of young and older women in an inner and an outer Melbourne suburb. Building on Valentine's work, Permezel confirmed that her respondents feared isolated places such as parks, streets with no houses, alleys, empty schools and streets, and confined spaces; isolation was especially feared in high-rise public housing estates, on trains, in multi-storey car parks, railway stations, lifts and empty lots. These women were least fearful in places where there were lots of people, and where there was a strong sense of neighbourhood. Women's fear was of being attacked, molested or abused by men. However, the gendered fear was complicated by issues of age—with older women feeling far more vulnerable than younger ones—and by race and ethnicity (Permezel 1992). These latter dimensions Permezel and Valentine only touch on. In Valentine's case, her respondents were all white and were particularly afraid of Afro-Caribbean men in certain parts of Reading (Valentine 1990). So too, in Springvale in Melbourne, Australian-born women had certain (undetailed) fears about Asian men (Permezel 1992). This complicating element of race was not explored further, and this is perhaps consistent with the radical feminist focus on male violence against women.

Permezel's work can be located within larger research projects in Australia investigating fear of crime. A national study commissioned by the Attorney General's department as part of a National Campaign Against Crime and Violence, found that public transport—including trains, buses and stations—were the most frequent source of fear, and that most people also have a general fear about unpredictable strangers, especially in public places at night. It also found that official crime statistics mask the constant low level of incidents which contribute to people's vulnerability and anxiety about becoming crime victims. The national study confirmed that women regard themselves as at greater risk of crime against their person, and that they feel unsafe in public places. Young women in particular are fearful of sexual and physical assault and are subject to continuous sexual harassment, especially on public transport (Crime Report 1998).

The responses—or coping strategies—detailed by Permezel for Melbourne and the Attorney-General's Department for Australia as a whole, were similar to those noted by Valentine in the United Kingdom and by Smith (1987) and Riger and Gordon in the USA (1981) and can be generalised as time–space avoidance. In practice, this means that women do not walk in public at night, that they go out in groups, use a private car where possible, are on constant alert for danger, check out safe locations and escape routes when walking or approaching a strange man, adjust dress and walk, carry weapons and secure their houses. Addressing such fears and actions, Valentine notes that men subtly insinuate or directly use the threat of sexual violence to exert power and hence control women's use of space. Given the existence of these sexual threats, Valentine connects feminist to geographical analyses (Pain 1991), and concludes that control over public space is achieved for all men (Valentine 1990, p. 301).

The 'cities of fear' literature, as one manifestation of the radical feminist position, has led to the rethinking and redesign of public spaces. In this way the radical feminist position has been inscribed onto space. While this research is not extensive, it has had widespread practical implications for women as recommendations are implemented by planners, community groups and urban designers.

The 'cities of fear' research is primarily concerned with the threat of male violence to women and with the abuse of heterosexual power; it does not usually examine violence against gay men, though the latter is an issue for those writing 'gay geography'. The policing of women in space is therefore part of a process whereby heterosexual relations are asserted and maintained and where a woman's chief strategy for avoiding danger is still the protection of a man. Those working in the 'cities of fear' area do not focus on sexuality, though sexuality is a concern in the radical feminist tradition. In the next part of this chapter, I explore compulsory heterosexuality in the Australian city in more detail in order to elaborate what a radical feminist geography could be.

SEXUALITY IN THE CITY AND SUBURBS

While the radical feminist agenda is primarily concerned with the definition and description of patriarchal power, its concern with personal–political issues such as fear, oppression and identity also embraced sexuality. Adrienne Rich's (1980) 'Compulsory Heterosexuality and Lesbian Existence' discussed the private and public mechanisms which prescribed certain sexualities as normative and others as 'deviant'. In the spirit of elaborating a radical feminist geography, the following section considers the ways in which a particular pattern of sexuality has been imposed on, or inscribed onto the Australian city in the nineteenth and twentieth centuries. It begins by looking at the nineteenth-century suburb as the outcome of culturally specific class and gender relations which also presumed and therefore helped to enforce heterosexuality. From these origins, the twentieth-century suburb was constructed as the quintessential heterosexual space: this occurred by government regulation (especially through public housing agencies), financial institutions, social expectations, house designs and neighbourhood layouts. Against this 'normal' environment is the inner city: the place of 'deviant' lifestyles and sexualities. The most obvious statements of queer or alternate sexualities in space are the Sydney Gay and Lesbian Mardi Gras, and gay precincts in inner Melbourne and Sydney. The following section argues, however, that despite the existence of such spaces, even the apparent freedom from prescriptive sexualities which the inner city offers occurs within a framework in which heterosexuality remains normative and dominant.

Nineteenth-century Suburbs

In his discussion of the nineteenth-century Australian city, Graeme Davison (1994) argues that four ideologies impelled and supported the quest for suburban life:

1 Romanticism—whereby the house in its own garden was deemed to be close to nature and thus a place of moral and physical regeneration
2 Sanitarianism—which saw the suburb as open, airy, cleaner, and spacious, and thus a healthier, less disease-ridden place than the inner city
3 Evangelicalism—a religious morality which extolled separate spheres for women and men and constructed the wife as the 'Angel of the Home'
4 Capitalism—whereby suburbs grew out of a physical separation of the paid workplace from the home, and were immensely profitable to create, own and sell.

Davison also notes how suburbs were counterpoised in popular discourse to the slum—to become 'the suburb of avoidance': consequently, the suburbs had connotations of social exclusiveness. Governments sanctioned and supported suburban housing through financial incentives and service provision; while suburban housing became increasingly accessible to the newly arrived migrant because of relatively high wages, cheap building materials, low unemployment, cheap land and extensive modern public transport (Davison 1994). Thus, nineteenth-century suburbs—as residential spaces physically removed from the city—were spatial statements of a particular cultural and class order. I would argue that underlying the cultural and class order was a specific gender order of heterosexual nuclear families, with home-based women and men in paid work.

Nineteenth-century Australian suburbs were derived from Imperial ideas about what was good for the social as well as individual body: these included property ownership, familial social relations, contact with nature, the country house, village life, fresh air, cleanliness, order, privacy and social exclusiveness; they excluded as undesirable renting, economic exchange, social anonymity, high density terrace housing, disorder, dirt and proximity to contagion, poverty, industry, the racially or ethnically different and the lower social orders. Critical to this ideological package was the placing of women and the home at the centre of an idealised domestic community separate from paid labour. In their essay 'Landscape with Figures', Davidoff, L'Esperance and Newby argued that British middle-class suburbia rose out of an idealisation of the rural village community and the home against the nineteenth-century industrial city–slum (Davidoff et al. 1976). In Australia, this suburban ideology was constructed as 'the quest for an organic community; small, self-sufficient and sharply defined from the outside world while the house and garden became the setting and symbol of the domestic community' (Dresser, quoted in Allport 1986, p. 236). In the United Kingdom and Australia, it was women who were entrusted with the creation and maintenance of that community, women who were to remove the dirt, restore the order and guard the privacy of this domain, and establish and maintain the culture of the suburb.

Australian suburbs also came to represent and accommodate an increasingly democratic class order. Initially the preserve of wealthy civil servants and professionals (as Sydney's Woolloomooloo was between 1820 and 1840 (Broadbent 1987)), the suburb later became the preferred location for the middle-class merchant and wealthy tradesperson. By the late nineteenth century, with the development of industrial suburbs, even those working in trades and factories could attain this form of housing. The embrace of the suburban ideal by the working classes modified the social exclusiveness of this built form. In 1891, the City of Melbourne still housed 15% of the urban population and the inner, terraced suburbs 39%, primarily of the working class; these were Collingwood, Fitzroy, Richmond, Carlton and South Melbourne (Howe 1994). Beyond these areas was a broad ring of suburbs differentiated by class and status. Thus 1890s Melbourne had a belt to the south and east of affluent, brick villa suburbs where owner occupation rates were high (as in Camberwell (Blainey 1980)); while Melbourne's north and west housed the industrial working class in wooden but still suburban housing, in suburbs such as Footscray, Brunswick and Northcote (Forster 1995; Lack 1991). The class division between suburbs was associated with particular gender relations: the physical separation between home and work which typified the middle-class suburb did not apply in the inner, northern and western suburbs; here the nature of jobs required workers to live close to their work, and low, irregular wages made public transport too expensive and ensured most of the household worked for money (Howe 1994).

In the middle-class suburbs, it was assumed that women would bear and rear children and supervise domestic labour in the home while the male breadwinner journeyed to the city for paid work. This public–private distinction was further enshrined in the social services and organisation of the nineteenth-century suburb: the church, the school and the shops, all of which relied on the patronage and labour of women to maintain them. This was the gender order on which the suburbs (and the city) were built. While initially a middle-class ideal, and the reality for much of the nineteenth century, it was also the model for all other groups who came to share in it. In these ways, the suburbs became important to the production and maintenance not only of a particular set of cultural mores and class relations but in specific roles for women and men in Australia. The remainder of this chapter will consider the way in which these roles were constructed especially as they came to be connected with a normative heterosexuality.

Heterosexuality and the Nuclear Family in the Nineteenth-century Suburb

The construction, marketing and occupancy of the nineteenth-century villa suburb made a set of assumptions which were enforced on the purchasers of suburban dwellings. The design of the houses, the layout of the neighbourhood and the operation of the city space economy, assumed that a married man and woman would purchase or rent the home. They also presumed that 'the man of the house' would leave it daily for paid work and the woman would remain behind to tend to the servants, servants being a reality for most middle-class women until the turn of the century. In the home and the backyard of middle-class households, servants performed a range of productive activities necessary for an expansive domestic economy; in working-class households, wives and unmarried daughters performed these tasks. This political and patriarchal economy was reflected in the design of houses and their surroundings.

The use of spatial layouts to 'read' social relations is a skill long practised by archaeologists and anthropologists and, more recently, by human geographers and planners (Huxley 1994). In architecture and the social sciences there are a number of studies which examine house plans to tease out gender relations and cultural assumptions about cleanliness, bodies, food, children and bathing in Western countries (see, for example, Hayden, 1981; Hourani 1990; Lawrence 1979; MacIntyre 1991; Matrix 1984; Roberts 1991; Rock et al. 1980; Spain 1992; Wright 1980).

In the USA, Daphne Spain (1992) and Rock, Torre and Wright (1980), have used house plans to read the gender relations of working-class, middle-class, and *ante-bellum* houses. They describe a set of designs deeply indebted to British precedents, especially the English gentleman's country house and the Victorian cottage both with rigid spatial and social separation—in a range of single-purpose rooms—of servant from master, women from men and children from adults. In these dwellings the front of the house contained the parlour where visitors were entertained and social status displayed; towards the rear was the kitchen. In two-storey houses the bedrooms were above living and work areas, with parents occupying the largest room at the front of the house and children having sex-segregated and private spaces of their own. To the rear or in separate outbuildings, were spaces for servants; these were also rigidly sex-segregated; in sum, a spatial statement of household and more general class and gender distinctions.

In nineteenth-century British houses this pattern varied between classes only in its scale and opulence. The Model House, built by industrial philanthropists such as Robert Owen and Titus Salt for their workforce, aimed to organise the working class in company towns into 'proper family units' in which there were to be no unrelated adults nor paid work activities (unlike earlier weavers' cottages). These houses were to be spatially separated from each other rather than built back to back or in long terraces; this would reduce the transmission of disease and instill notions of privacy. Such houses have persisted in the form of twentieth-century council housing (Matrix 1984). The urban terrace, while joined, was little different in conception, with the kitchen, bathroom and scullery at the rear, parlour at the front and bedrooms upstairs (Matrix 1984).

In the United Kingdom, the Victorian gentleman's city house was pervaded by modest elegance, extreme propriety in personal behaviour and sharp distinctions between master and servant, and between women and men and adults and children (to the point where there could be separate entrances and stairways for each group). In these dwellings, status decreased from front to back and with greater distance from the main floors. Work areas were efficient, functional and hidden, while entertainment spaces were grand and at the front of the dwelling. Eating was separated from living areas, and bedrooms far removed from both. English, but not US, city houses of this era, contained a separate scullery, where washing was separate from the preparation and eating of food in the kitchen (Lawrence 1979).

In Australia, the parallels are remarkable and point to the importance of the Imperial legacy of form and ideology, as well as to the proliferation of North American design ideas and domestic technologies. However, a few significant differences prevail: these include the preference for detached over attached housing, styles of façade and scale. In addition, the placement of the kitchen and scullery, and the size and use of the back garden differ from their overseas counterparts. These differences point to particular constitutions of class and gender relations in Australia. As Rock, Torre and Wright note: 'there is a close correlation between the way houses have been designed and socially sanctioned ideas of sex roles and domesticity' (1980, p. 93). These ideas will be developed drawing on a sample of house plans from eastern Australia from the 1790s, 1850s and 1880s.

The first house plan is from a sketch done in 1795 on Norfolk Island; it shows that the earliest white Australian dwellings retained the British ideal of a formal parlour at the front of the dwelling and a kitchen separate from the house. This pattern persists into the 1850s (see figure 3.1) when, despite the obvious increase in the scale and opulence of the dwelling, the front continues to be occupied by the formal spaces of the parlour which is reserved for 'women's activities' and the drawing room for men's. The bedrooms are towards the rear with the kitchen still at the remotest point in the house. These plans and those from the 1880s, show that, regardless of the class of the occupiers, the status pattern of rooms and functions is retained. Greater affluence means more rooms, and also spaces where the distinction between servant and master could be made more apparent; as could the distinction between parents and children.

While this level of analysis and its evidence prohibits firm assertions about how these spaces were used, and the meanings they may have had for their occupants, it is still possible to offer some credible observations on the social relations suggested by this layout. In the smallest houses of the 1850s there were few gender differentiated spaces; however, in the sample of larger houses from the 1880s, it is clear that it was expected that their occupants would be a heterosexual couple and their children (see figure 3.1). The plans attest to women's labour, which was performed either by servants or by the 'housewife', being done in buildings or rooms that were physically removed from, or at the rear of, the main part of the house. At the front of the houses is the parlour, sitting or drawing room, a formal space reserved for guests and leisure. There is invariably one large bedroom, and in the smallest dwellings this might have accommodated the couple as well as their children; beyond that were other, much smaller bedrooms towards the rear of the house; these accommodated the lower status and sex-segregated children. Towards the rear of the house are the work spaces for food preparation and spaces for bodily care: toilets and bath-rooms, their placement perhaps indicating the taboos, and the private and low value placed on these activities as well as those usually associated with them, that is, servants and women.

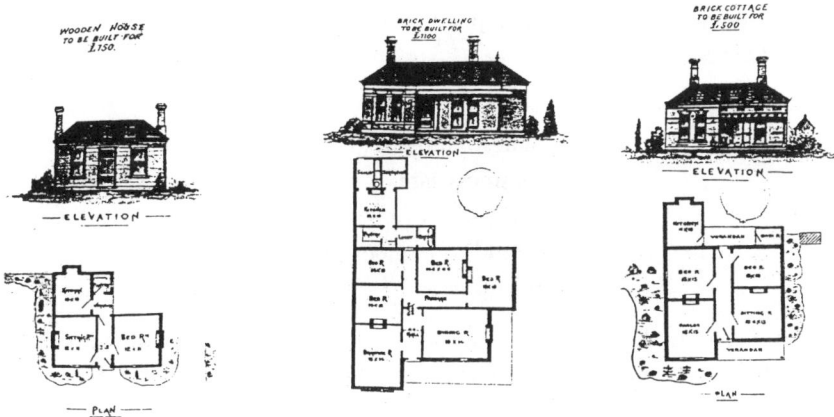

Some contemporary house plans and elevations (1885)

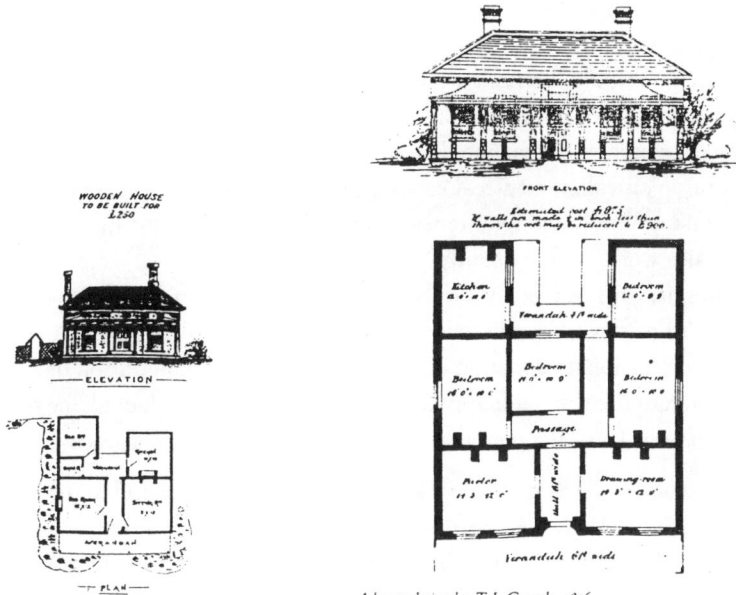

A house design by T.J. Crouch, 1856.

Figure 3.1 House plans in nineteenth-century Australia
(Source: Davison 1978)

These house plans reveal class and gender distinctions, yet public statements emphasised only one aspect of these homes: the gendered physical separation of home from paid work; such statements highlight the importance of this dimension in the nineteenth-century conceptualisation of suburban living. Thus

Alexander Sutherland who compiled *Victoria and its Metropolis* wrote in his poem 'Home and the World':

> So be thou in the world. So raise thy life
> On the one side swelling from the petty strife
> Of men and business care;
> But on the other, where
> Thy Home extends the smoothness of its breast
> Sinking in trustful rest.

In 1890, a Mrs T. Harris drew out the connection between the woman's place in the home and that of men in the public world of commerce in her *Woman— the Angel of the Home and the Saviour of the World*. Here she writes of the home as the husband's paradise, his refuge, his pride, his castle, but also the place where the wife rules. It was to this world that the husband returned after a busy day in the world of commerce to be met by 'happy greetings by wife and children (after) his struggle with the contending currents of the world's seething sea without'. Then he would 'lounge', attend to the garden or play with his children (cited in Davison 1978, pp. 137–9).

While such statements emanated from those writing about the late nineteenth-century Australian city, real estate promotions for suburban estates also touted their naturalness, and remoteness, emphasising the contrast between the world of paid work and womanly domesticity. These views were widely expressed despite this era being one where large numbers of (working-class) women were in paid work, and successful campaigns were being waged for the vote and the professional education of women. However, such agitation did not necessarily undermine the public/private distinction and, indeed, could be seen as building upon it (see chapter 1), as the following comments suggest:

> There are women to whom the pursuit of pleasure is not the end and aim of existence, who desire something more than to wear pretty toilettes and to be admired...and this is one class to whom the Women's College will confer a boon [but] The Women's College is not likely to destroy the domestic future of the girls of New South Wales. It will be no fashionable craze to interfere with home life (Florence Walsh 1890, cited in Johnson 1984, p. 68)

The construction of the suburb as a place of retreat for men, and of homely and mothering virtue for women, occurred against the backdrop of the inner city slum, a place where women as well as men worked and where the threat of social disorder emanated from the crowded tenements of the poor. The inner city was also the place for the socially deviant—the prostitute, the dispossessed, the Chinese—a deviation which highlighted the ways in which the suburb was a place for the Anglo-Celtic nuclear family. In the twentieth century, the contrast between the inner city and the suburb was similarly constructed, though to this array of familial, economic and ethnic contrasts was added an explicit articulation of the differences around sexuality. Thus the contemporary suburb is the place where heterosexuality is explicitly affirmed against the sexual deviance of the inner city.

Sexuality and Space in the Twentieth-century Suburb

The beginning of the twentieth century heralded significant changes in the division of labour which had underpinned the Victorian household. The decline in the availability of domestic servants—who had all but gone as a mass profession by the 1920s—and the fall in the proportion of women in the labour force over the first half of the twentieth century, meant that women's roles in the home underwent major redefinition. The middle-class woman, no longer the overseer of labour, increasingly had to do much of the domestic work. The working-class woman, no longer employed as the paid domestic servant and increasingly banished from paid work, also had to retreat to her home. In the process, the nature and meaning of housework changed from being recognised and remunerated labour to a scientific activity engaged in by a devoted wife, family servant and highly skilled home maker (Matthews 1984). This new found status for women's domestic labour was reflected in the creation of the Housewives Association and the Country Women's Association. The houses in which such work was to be performed became far more focused on family inter-action around what was now the core of the house, the kitchen, which moved from being either a separate building or at the rear of the house, to being at its centre (Hourani 1990).

In houses in the USA and the United Kingdom in the 1920s, the parlour all but disappears, to be replaced by the living room, a multi-purpose room for all the family. By the 1940s in the USA, most rooms were no longer single-purpose but served multiple purposes. With the growing popularity of the horizontally spread 'Ranch House', spaces flowed from one part of the house to another. The open plan had arrived. But not in British house designs, where spaces were still more segregated and the emphasis was on modern technology in kitchens and bathrooms. In the British Garden City New Towns built after the Second World War, light and airy houses were located in well-serviced semi-rural neighbour-hoods. Here experiments also occurred in the design and use of domestic space, with an attempt, during and immediately after the war, at shifting cooking and laundry tasks out of the home and into the public sphere. Alongside such devel-opments went an idealisation of family life and of the home-based woman

(Matrix 1984). Such a pro-family ideology, which did not challenge domestic labour arrangements, was pursued with renewed vigour, along with the drive to own suburban housing in Australian model suburbs such as Sydney's 1920s Daceyville (Freestone 1989; Hoskins 1994).

In Australia home ownership has long been associated with suburban living, and with a particular class ideal and gender order. Thus, in the late nineteenth century, inner Melbourne city areas such as Richmond, Collingwood, Carlton and South Melbourne had rates of home ownership of around 20% to 30%. In contrast, the new suburbs of Hawthorn, Kew, Northcote and Footscray had ownership rates of between 40% and 60%. In the early twentieth century, the rate of home ownership was relatively stable at around 45% across the country. However, since the Great Depression and the Second World War, there have been attempts encouraged by Federal Governments to raise this level, in partic- ular to give workers and their families a 'stake in the country' through suburban home ownership.

In 1950s Australia, the emphasis in public discourse was on women leaving the paid workforce, where they were actively recruited during the war, and returning to their homes to have babies. In the domestic arena, women were also expected to rebuild their war weary men and help create a stable, affluent and conservative nation in the face of the threat of Communism. Prime Minister Robert Menzies, in his famous 'Forgotten People' speech of 1942, asserted: 'One of the best instincts in us is that which induces us to have a little piece of earth with a house and garden which is ours, to which we can withdraw, in which we can be amongst our friends, into which no stranger can come against our will' (quoted in Davison, Dingle and O'Hanlon 1995, pp. 4–5).

These views underlay Federal Government policies to increase the supply of new houses and to lower the cost of borrowing money for new home purchase through a variety of mechanisms: these included War Service Loans, lump sum cash payments for service personnel and an easing of restrictions on cooperative societies and banks. The activities of State and Federal public housing authori- ties, together with undermining of rental controls and an expansion in consumer capitalism, meant that both home ownership and new, outer suburban housing grew rapidly at the expense of the inner city and the rental market (Allport 1986; Game and Pringle 1979; Kass 1987). In Sydney, for instance, between 1947 and 1954, owner-occupied housing rose from 39% to 56% and, again, to 71% in 1966 before it began to fall. Between 1947 and 1961, the 38% of the city's population who lived in the established inner city and eastern suburbs decreased to 25% (Allport 1987; Kass 1987). Thus, as Kemeny argues, home ownership (and mass suburbanisation) was not something which just happened naturally in Australia. The growth in home ownership rates was a result of conscious State policies, effected through home finance, tax exemptions and direct grants, as well as actions to weaken the private rental and public housing sectors (Kemeny 1981, 1988). These policies both encouraged home ownership and the nuclear family, and linked them as part of a conservative Cold War agenda to ensure political stability, male working-class loyalty to their jobs, their family and nation, female fecundity and domesticity.

Both the lending practices of banks and the allocation policies of government housing authorities explicitly reinforced the suburban heterosexuality implicit in the rhetoric about housing the nation. As Carolyn Allport observed:

> In the immediate post-war period the institutional approach to housing assumed that women exercised their access through marriage...and the family (including service widowhood)...There is no mention of single women at all, even if those single women had dependants...Similarly, suitability for a war-service loan...necessitated that the applicant, whether serviceman or servicewoman, should satisfy the director 'that he is married or is about to marry, or has dependents for whom it is necessary for him to maintain a home'...Further...the Federal government favoured the preferential granting of home loans to families with children; although it was never actually expressed as a distinct policy from the central bank, it was certainly part of the political rhetoric throughout the 1950s and early 1960s (Allport 1986, pp. 239–40).

The result of these policies can be seen in the many public housing suburbs built to house working-class and immigrant families in the 1950s, 1960s and 1970s; examples include South Australia's Elizabeth (Peel 1995), Melbourne's Doveton (Bryson and Thompson 1972; Howe 1995; Bryson and Winter 1999), and Sydney's Mt Druitt (Dowling 1996), as well as many private developments which made up the expansive suburban frontier. From 1945 to 1955 over 700 000 houses were built in Australia, an increase in ten years of 25% in the nation's housing stock. Of these houses, public authorities built one in seven, whereas owners built one in three. All were designed for the nuclear, heterosexual family.

The houses built at the time reveal the gender relations assumed and prescribed within them. Figure 3.2 illustrates the first in a series of Federal government designed houses, a unique development in the history of the Australian suburb which, up to this point, had been built primarily by private

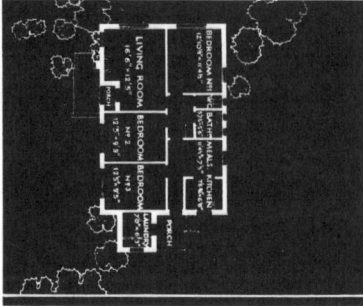

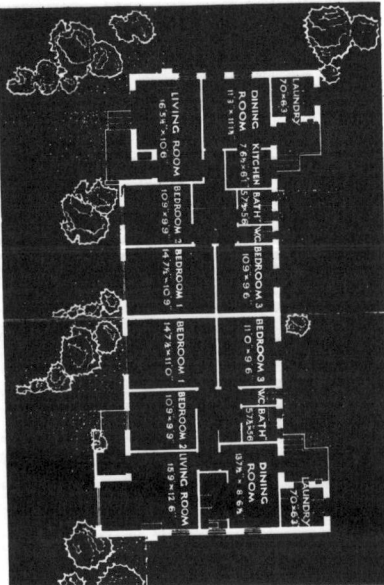

Figure 3.2 Australian government house plans from the 1940s
Source: *Australian Housing Bulletin* 12, 1947

TYPE B75

Construction: Brick.

Accommodation:
Three bedrooms,
Six persons.

Gross Area: 1,098 sq. ft.

Total Net Floor Area:
929 sq. ft.

TYPE B70
(Opposite page)

Construction: Brick

Accommodation:
(each unit):
Three bedrooms,
Six persons.

Gross Area of Four:
2,183 sq. ft.

Total Net Floor Area:
1,841 sq. ft.

Architects: The Housing Commission of New South Wales

330

331

interests. In response to massive material and housing shortages after the war but also inspired by visions of building a fairer and richer nation, these dwellings were humble in scale and were made widely available to the 'right' kind of (nuclear) family. They had a number of characteristics which made them dramatically different from the houses built at the end of the nineteenth century: their horizontal placement across the block (reminiscent of the US Ranch House), the existence and purpose of the living room, and the placement of the kitchen adjacent to a dining area. As befitting the changes in the status of domestic labour and the women who performed it, the kitchen has moved towards the centre of the house and is exposed, being linked visually and phys-ically to an open living area. The formality of the parlour has given way to the informality and family-centredness of the living room.

However, the status of the backyard remains similar to that of the nine-teenth-century dwelling. Until the 1970s, the yard—as distinct from the garden—was entered through a small doorway at the back of the house: it was a work space, a place variously occupied by bodily or noxious functions (the toilet, laundry, clothes drying, kitchen, stables), or by those activities necessary for the economic survival of the household, such as fruit trees, vegetable garden and chicken coop. After the Second World War it was also occupied by the Hills Hoist and the garage. The yard was not a space to view or one into which guests would be invited; rather it was a work space for both men and women (Fiske, Hodge and Turner 1987; Seddon 1991; Johnson 1999).

Suburban houses were placed in neighbourhoods where the rhythms also reflected the dominant gender order. Thus in 1950s and 1960s suburbs, men tended to live their lives at a regional scale: they took the family car to large industrial plants, to drive in hotels and to sporting venues. In contrast, women were often unable to drive the one car in the family and were bound to the home by young children: thus their lives remained decisively shaped by the more intimate and spatially concentrated relationships between home, local school and shops (see chapter 4), and by employment to which they either walked or journeyed on public transport (Davison and Dingle 1995; Frost and Dingle 1995; Spearritt 1995).

Analysing the place of women and housing in post-war Australia, Sophie Watson highlighted what she defines as its underpinning 'family discourse'. She elaborates:

Pride in the family and pride in home ownership are presented as related notions; the one enhancing the other. It is the *patriarchal family* form which is idealised and promoted: man as the protector, the benefactor, the provider; such images are abundant. [For example, in debate] on the first

Commonwealth–State Housing Agreement in 1945...'The safety of this country...depends on private ownership which gives a man a stake in the country, a place of his own in which to keep his wife and rear a family, and a place for his family to live when he dies' (Watson 1988, p. 7) (my emphasis).

Women in this era, in contrast, are constructed as dependants, as 'housewives', and as 'homemakers'. But, as Watson notes, it is not only at the level of discourse that home ownership reproduces a dominant gender order and sexuality; rather the structure of home ownership also acts to exclude many women and reinforce their dependence on male partners. This occurs in two main ways, (i) through women's place in the labour market and (ii) the attitude of lending institutions.

According to Watson, it has only been since the 1960s that more women have entered the paid workforce and have earned an independent income to qualify them for home finance. From their earliest participation in the paid workforce, though, women's wage levels have generally been far lower than men's. This followed the Harvester Judgement of 1907: the ruling that the basic female wage should be 54% of the male rate because it was assumed that the man would be the breadwinner (see chapter 1) and that a sexual division of labour was part of the labour market. Thus women tend to work in the sales, clerical and personal service areas at relatively low levels and often part-time (see chapter 2). Hence, despite over 30 years of Equal Pay legislation, women continue to earn lower incomes than men, while the occupations in which women work mean that they find it difficult to raise a house deposit and to satisfy lending institutions of their ability to repay large amounts over long periods of time.

Watson suggests, secondly, that the practice of lending institutions has discouraged women from becoming home owners. For example, the 1964 *Home Savings Grant Act* only made subsidies for home purchase available for married couples aged under thirty-six with dependent children. In Queensland in 1978, eligibility for home purchase assistance under the Housing Assistance Act was for families only, and in South Australia for parent(s) with dependent children and couples aged under thirty without dependants. While these schemes now include de facto couples and single people, earlier regulation by private and state lending agencies was part of the massive edifice which ensured that home ownership and suburban living was accessed primarily by heterosexual couples.

Watson also reports cases of discrimination against women: in 1981 investi-gations into NSW finance companies showed that these companies preferred to grant home loans to a woman when she was prepared to be sterilised! A more recent survey of female applicants for loan finance revealed cases where women were confronted with chauvinistic and patronising attitudes, even harassment from the staff of financial institutions. Likewise, women applying for loans with male partners reported that their income was not taken into account, suggesting assumptions were being made about their future employment, the likelihood of their future pregnancy and their automatic withdrawal from the labour force during the early years of motherhood. These assumptions about women are not made about men: namely, that women will leave the workforce to have children, that they will be unreliable in maintaining loan repayments and that they should therefore not have access to finance unless a man stands guarantor (Watson 1988).

The consequences of such assumptions and practices over the 1950s, 1960s and 1970s are variable rates of home ownership for different social groups. Thus as table 3.1 indicates, in 1981, 75% of nuclear family households with dependent children and with a male head of household owned or were purchasing their dwellings compared with 46% of female, single parents with dependent children. So too, for house sizes. A higher proportion of traditional nuclear families who are owners–purchasers live in separate dwellings than do single women with children. Statistics on dwellings with five or fewer rooms, show that 25% of nuclear family households live in these dwellings compared with 37% of female, single-parent families (Watson 1988, p. 31). In contrast, women—especially single parents—are over-represented in public housing and in the private rental sector. Indeed, the only time when women as individuals secure home ownership in significant numbers is in old age when their spouse dies and leaves them the 'family home'.

In this period from 1945 to 1981, shifts in the gender order have also been registered in house designs. Compared with the 1940s, the 1960s and 1970s have emphasised individual family members having their own space in multi-purpose spaces; this applies especially in 'family' or 'living areas', although the kitchen remains at the centre with the housewife bridging the various living, eating and working zones of the house as well as bringing the family members together. From the 1960s, when houses began to be unoccupied during the day as married women started to work outside the home, the trend has been towards the further opening up and multiple use of spaces (such as the main bedroom flowing into its adjoining bathroom and the recreation room flowing outside), greater size, less formality and less separation of family members and women from men (Spain 1992).

Family type	Couple		Couple and dependants		Head and dependants		Single household	
	Male head	Female head	Male head	Female head	Male head	Female head	Male head	Female head
Owner	48	37	23	17	23	16	27	43†
Purchaser	28	31	52	50	38	30	17	12
Public renter	4	4	6	8	6	20	4	8
Private renter	12	19	12	17	27	27	38	26

Table 3.1 Housing tenure* by sex and family type, Australia 1981**

* Figures derived from the 1% household sample file of the 1981 census.
** Figures are not available from the 1991 and 1996 censuses.
† Mostly pensioners
Source: adapted from Watson 1988, p. 43.

Comparing house designs from the 1960s and the 1980s in Melbourne and Perth, the architect Kim Dovey describes a number of differences, the main ones being an increase in the overall segmentation of the house, a rise in the size of dwellings (from 15 squares in 1970 to 22 in 1989), and an increase in the importance of the interior of the house which is now connected to a courtyard or a formal outdoor living area (Dovey 1994, pp. 128–40) (see also Garden 1995). Dovey notes the changes:

- a shift in sleeping areas away from a central hallway to zoned areas for parents and children in different parts of the house
- the parents' area has been enlarged and elaborated to include separate bathrooms and separate living areas
- a change from a small kitchen-meals area to an expansive kitchen-meals-family-games area where most informal family and entertainment activities occur (while formal dining-lounge spaces remain about the same size)
- the backyard has moved from being a poorly accessible work space to two spaces, one flowing from the inside living area into an overall entertainment space and the other being a hidden service area.

Dovey sees many of these changes as reflecting different gender relations as women move from being the efficient kitchen manager and child overseer to entertainer, chef, paid executive and child carer. He also sees women portrayed in home advertisements less as passive occupiers of housing and more the decisive agents in the use of the home. However, despite these changes, Dovey acknowledges that 'patriarchal assumptions persist', as the wife and the house assume the role of integrated symbolic package to bolster male esteem, privilege and economic position. Women still do most of the domestic labour even when they are in the paid workforce: 'While the position of women in public life in Australia has changed markedly, it could be argued that the home is a haven for men from the public demands for equal opportunity' (1994, p. 140).

In sum, these developments reveal continuity in normative gender relations and their reflection in house designs, as well as the very real changes in women's positions and in the house designs offered to them and their households. In the 1970s, two developments disrupted the standard home/workplace split of post-war suburbia. These were the suburbanisation of white collar service and clerical jobs—especially in part-time work for women, and in full-time manufacturing jobs for male (often immigrant) workers—and the expansion in paid home work (Fincher 1993). Further, the nature of the family, so long extolled and supported by the financial institutions, builders and public housing authorities, also began to alter. As table 3.2 indicates, the nuclear family, of a couple with children, is increasingly a minority household form, with its numbers shrinking relative to the couple with no children, the single person household and 'other family' forms. Consequently, the New South Wales Housing Commission, for example, in 1974 made housing available to divorcees and, by the mid-1980s, most of their new tenants were single women and their children. Developers too, are responding to the growing diversity in household types, by designing houses not just for the nuclear family but houses on smaller blocks suitable for the child-free couple, 'empty nesters' whose children have left, and the wealthy, single individual.

Family type	1986	1992	% change
	000s	000s	
Couple	3507.9*	3835.0	+9.37
with children	1548.7	1608.0	+3.8
without children	1959.2	2227.0	+13.7
One-parent families	270.2	340.7	+26.1
female parent	239.0	309.8	+29.6
male parent	31.2	30.9	−1.2
Other families	281.9	348.8	+23.7
Total families**	4060.0	4524.5	+11.4

* Corrected to nearest full number
** Single member households are not considered a family

Table 3.2 Family types: numbers and proportionate change, Australia 1986 and 1992
Source: ABS 1993, p.18; 1998 personal communication

As economic restructuring occurred, and unemployment and interest rates rose in the 1980s, the rate of home ownership fell and Australian cities, and the class structure, bifurcated. While the inner city underwent a revival the suburbs were increasingly occupied by poor, non-nuclear families, single people and the elderly. In the 1990s, it seems the 1950s and 1960s rigid regime of sexual regulation is disappearing and being replaced by openness, tolerance, even a market for, and social acceptance of, a range of alternative family forms and practices. The postmodern turn towards sexual diversity and play appears to be mirrored in the contemporary Australian city, even, perhaps, in its suburbs. Such an assertion, while partly true, underplays the continued assumption and enforcement of a normative heterosexuality in the suburbs.

Contemporary Suburbs

Recent research into who lives in the outer suburbs of Australian cities has raised questions about whether this is where nuclear families buy their first homes. The national Housing and Locational Choice Survey (HALCS) conducted for the National Housing Strategy in 1992, found that in fringe suburbs, those living in their first purchased home were outnumbered by changeover buyers, that is, those who had previously owned a home. Forty per cent of households on the fringe were first-home buyers, 44% were changeover buyers, while 15% were renting (Burgess and Skeltys 1992, p. 10).

A range of house and block sizes are emerging in the outer suburbs of large Australian cities: thus developers are recognising and meeting the market demand from new household groups, especially for the single person, the child-free couple, the older couple, the empty nesters, the blended family and so on. Similarly, lending policies and government regulations on access to home ownership have increasingly recognised the single woman and her children as an entity with particular housing needs, and as headed by someone able to afford repayments.

Thus it seems that the suburb is no longer just the place for nuclear families. However, it is still the place where, despite the recognition of difference, heterosexuality is regarded as the norm. Patriarchal relations and notions of compulsory heterosexuality can be readily seen even in the most recent house designs. Figure 3.3 shows block arrangements offered by the Delfin Property Group for a new housing estate in Adelaide. One of the largest and most progressive development companies, the Delfin Property Group offers a range of house designs and plot sizes which engage with the greater social diversity of the suburbs (Johnson 1997).

Thus at Mawson Lakes twelve kilometres north east of Adelaide, as at the other fifteen Delfin developments across the country, there is a wide choice between Town Cottage, Villa, Patio, Carriageway, Courtyard, Cottage and Traditional block and house designs. All but the Traditional are less than 600 square metres; this compares with the standard quarter acre lot, and addresses households other than 'growing families' which, in the past, have been the main market for suburban housing. Requiring blocks from 270 to 420 square metres

RADICAL FEMINIST GEOGRAPHY 109

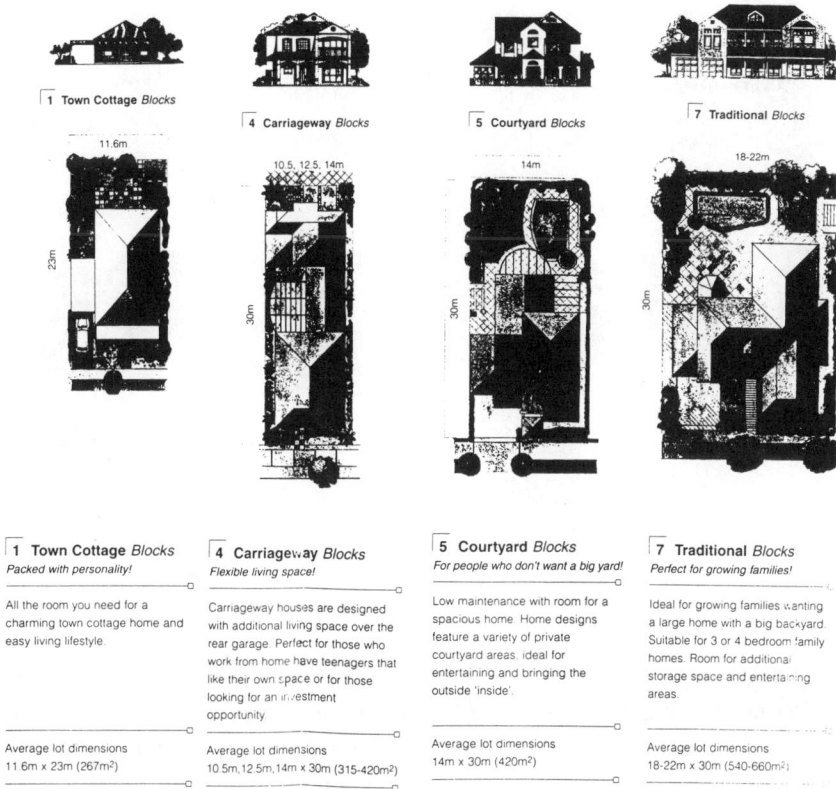

Figure 3.3 Housing and block arrangements offered by the Delfin Property Group at Mawson Lakes
Source: Delfin Realty n.d. Mawson Lakes

and with houses from 16 to 20 squares, the Town Cottage, Villa, Courtyard and other designs are aimed at particular household configurations, those of the single person, young first-home owning couple, small, young family, family with teenagers, mature couple and retiree: as the HALCS survey confirmed, these groups are present on the urban fringe (Burgess and Skeltys, 1992). However, as the promotional material makes clear and as the photographs of happy occupiers attest, these social groups are not alternatives to the heterosexual norm. As figure 3.4 indicates, the smiling couples and those with children suggest in their warm embraces, that these are not houses or suburbs for gay couples, single mothers or group households of unrelated adults. Heterosexuality remains the norm in the suburbs for those who design them, and in the marketing to those who are expected to occupy them.

Figure 3.4 Residents of Delfin homes, Mawson Lakes
Source: *Mawson Lakes Update*, Issue 4, November/December 1998

This radical feminist geography of the twentieth-century suburb has insights and lacunae. It highlights the solid foundations of government, financial institution, planning and marketing regulations and promotions which have made the Australian suburb a place of normative heterosexuality in different ways over time, but it does not necessarily indicate what occurs in the suburbs or their houses. The argument is not that the spaces of the house somehow determine what occurs in them, nor that government and financial institutions through their regulatory statements and practices generate total compliance. A radical feminist geography framework merely asserts that dominant patriarchal and heterosexual relations are inscribed and enforced in our urban environments in a number of ways which can be isolated and which have very real material effects. This prescribing is most clearly seen in the creation and recreation of the Australian suburb. While the suburb has long been counterpoised to the inner city as its civilised, familial and urbane other, I would argue that the so-called gay city is also prescribed by the various practices which maintain patriarchy and compulsory heterosexuality as dominant social relations.

Gay Cities

In contrast to the idyllic notion of the postwar suburb, the inner city has long been demonised as the place to avoid, for its dirt, disease, high density housing, poverty and the working class. Since the Second World War, the city has also been host to the racialised and sexualised 'Other': accommodating the gay community and migrants from Europe; this devalued area offered cheap housing for the newly arrived immigrant; housing various 'Red Light Districts' it was a place where single women with children could afford to live and work. The inner city was also one site for gay men to meet, live and socialise (Wotherspoon 1991; Aldrich and Wotherspoon 1992).

Gay geography has examined how communities, identity and experience are constructed, and it details the ways in which sexuality affects these communities. Thus, for example, in his study of San Francisco, Manual Castells identified an inner city, gay urban territory; he drew on key informants from that community, multiple male households on voter registration files, the presence of gay bars and other social gathering places, gay businesses, stores and professional offices and votes for a gay political candidate (Castells 1983). Castells concluded that the predominance of homosexual men in a distinct gay neighbourhood reflected a profound gender difference; where men sought to dominate and delimit space while women were more scattered and had more non-localised relationships and networks. Similarly, Micky Lauria and Larry Knopp have used the presence of gay businesses as key indicators of the class, as well as the gendered power of gay men to create urban territories as their own, powers which women, being generally of a lower socio-economic position, cannot wield (Lauria and Knopp 1985; Knopp 1990, 1992). This economic power and the tendency of gay men to congregate in the inner city has meant that they have played a critical role in the gentrification of whole neighbourhoods in the USA and Australia (Ingram et al. 1997).

This tendency is spectacularly evident around Taylors Square in Sydney, and with the Gay and Lesbian Mardi Gras. The largest generator of international tourist interest and dollars, this Mardi Gras involves a dramatic identification and celebration of gay inner city Sydney (Murphy and Watson 1997). The identification extends to the shopping areas around Taylors Square and the adjacent suburbs of Paddington and Surrey Hills. As much gay geography has noted, the inner city precincts of San Francisco, Minneapolis, New Orleans and Sydney have become centres for male homosexual investment, business, housing and lifestyles (see Adler and Brenner 1992; Castells 1983; Knopp 1992; Lauria and Knopp 1985). Such activity marks off the inner city as gay and the suburbs as quintessentially straight.

Nevertheless, this distinction has not been empirically validated. Australian censuses before 1996 have not asked about same sex households and there have been no accurate measures of the extent of gay or lesbian residence in suburban versus inner city areas. Further, despite the media hype surrounding the activities primarily of gay men in inner Sydney, there is research occurring on gay communities in the western suburbs of Sydney (Hodge 1995, 1996), and research in the United Kingdom which suggests that lesbian communities are far more likely to be scattered across space and linked by networks rather than concentrated spatially in particular parts of the city (Valentine 1993a and b).

The inner city is also no nirvana of diversity, no place for the uninhibited celebration of social, sexual and ethnic difference. Rather, it is a place of homophobic violence and forthright police regulation of 'deviant' sexualities. Just as the suburbs are associated with a dominant sexuality, so too a range of activities in the inner city reinforces heterosexuality. Stephen Hodge and Stewart Kirby highlight some of the ways in which heterosexuality is enforced to limit the expression of gay male sexuality in the inner city as much as in the suburbs. Thus Kirby notes how even the homes of his gay male respondents—that quintessential site of individual expression and privacy—are variously 'straightened up' for the gaze of visiting and disapproving parents, friends and colleagues. Kirby further observes that public spaces are also rigorously policed for any expression of what is clearly unacceptable 'deviant' sexuality (even in the centre of Adelaide); and that this occurs primarily by the threat and reality of physical violence (Kirby 1996).

Gay men whom Kirby interviewed lived in a range of different environments, some alone, others with partners, or in shared households, and with parents; Kirby does not tell us, however, which interviewees lived in inner or outer Adelaide. His respondents saw the inner city as a place where there was greater social diversity (especially in age and household variability compared to the suburban nuclear family) and hence safety for gay men. It was also the site for recreational activity which involved visibility, crowds and drinking, and hence often volatile combinations which could lead to anti-gay violence (Kirby 1996).

The inner city's danger (as well as its relative safety) is a point echoed by Stephen Hodge who discusses the prospect of a group of openly homosexual men walking between two of Sydney's better known inner urban gay havens, those of Darlinghurst and Newtown; his interest is in highlighting the fluidity, vulnerability and relativity of gay spaces in the city. Hodge tells the story of the anxiety which accompanied the walk, a sense of panic which enveloped members of the group at various locations in their traverse. Those walking could not agree on the point at which the safe spaces ended and the risky ones began, indicating again that the security of the inner city, even at the time of Sydney's much lauded Gay and Lesbian Mardi Gras, is transitory, tentative, and relative (Hodge 1996). Nevertheless, as Hodge has noted elsewhere, gay men are certainly not confined to inner city Sydney. The need to recognise the presence of gay men in the western suburbs of Sydney necessitated a reorientation in AIDS education strategy, which is usually directed in line with the notion of the inner city as the gay city. The recognition of suburban gay male communities also impelled the organisers of the Gay and Lesbian film festival in 1997 to have films running at both inner city Paddington and in the western suburb of Parramatta (Hodge 1995). The point here is that the suburbs are not only home to heterosexual nuclear families—as many anecdotes· confirm, gay men and women live there too—but rather that the inner city is not the only haven for gay men and homosexual culture. Furthermore, there is regulation of that sexuality at both sites.

The issue of how sexuality is defined and lived is one which has been explored in another body of theory informing feminist geography, that of Postmodernism. Postmodern feminists have questioned the simple dichotomy between male and female and the centring of male power in a concept like patriarchy which underpins radical feminism. The next chapter examines their work and its applicability to the study of regional shopping centres.

■

4

Postmodern Feminist Geographies

What is postmodernism? Postmodernism is highly contested and feminists' attitude to the notion ranges from outright rejection to ambivalence and critical engagement (Nicholson 1990; Singer 1992). Postmodern geography has welcomed its concepts and insights—especially in relation to the contemporary economy and city—but failed to recognise the place of women in postmodernism. Feminist geographers have been eloquent in exposing the ways in which postmodern geographies have variously excluded women and feminism, while also recognising the possibilities for new feminist geographies that feminist postmodernism and deconstruction offer. In this chapter, I attempt to capture the diversity and contested nature of postmodern feminist geography, and, using one example, to show some of the insights from this dialogue by interrogating a contemporary landscape. This is a regional shopping centre in Australia's global city and centre of spectacle, the city of Sydney.

First, I give an overview of postmodern geographies and discuss some of the critiques from feminist geographers. I then discuss some postmodern feminist geography to outline the terms of this engagement and its tools and analytical directions before using this framework to examine Westfield, a shopping centre in Parramatta, western Sydney.

POSTMODERN GEOGRAPHIES

Geographers and other social theorists see the postmodern primarily in three ways (after Dear 1986):

- an era when contemporary capitalism globalised and changed its form from a Fordist form of mass production into a disorganised, flexible or post-Fordist way of structuring production, primarily in the service sector rather than in manufacturing plants
- a style, a way of organising architectural and design elements in space, in which buildings are both double coded—with modern, local, historical, and vernacular or eclectic elements coming together in a play of paradox, pastiche and parody—and fortified with walls, gates and security
- a crisis of representation for general theoretical frameworks and accepted truths drawn from the Enlightenment.

Geographers have primarily focused on the first dimension, taken good note of the second, and acknowledged the third but have not generally incorporated it into mainstream geography. Feminist geographers (as well as 'cultural and post-colonial geographers', see chapter 5) have paid more attention to the third dimension. For many feminists, existing postmodern geographies are sexist, have marginalised the place of feminist scholarship and ignored the profound contribution a feminist perspective can make to the analysis of postmodern landscapes and conditions. For feminist geography the crisis of representation has mostly meant exposing the patriarchal politics of writing within modernist and postmodern geography. It is from this foundation—as well as from a number of other postmodern concerns such as deconstruction, the body in space and multiple identities—that feminist geography has drawn most energy. These elements inform my study of Westfield Parramatta. First, however, I locate this regional shopping centre within Sydney's post-industrial, post-Fordist economy.

Post-Fordism—New Times for Capitalism

For Frederic Jameson, postmodernism formed what he saw as 'The cultural logic of late capitalism' (1984). In this logic, the postmodern involves the prodigious expansion of capital into hitherto uncommodified areas of activity under the auspices of US multinational corporations and a decentred global communications network. Others have described these tendencies as post-industrial (Bell 1973), disorganised capitalism (Lash and Urry 1987), post-Fordist (Cooke 1988; Mathews 1989; Piore and Sabel 1984), New Times (Hebdidge 1989) and flexible accumulation (Harvey 1989a and b). These formulations focus on the increased globalisation of corporate capital, the way it is organised and what is produced.

In these narratives of economic transformation, a comparison is drawn between the mass production of standardised items for large undifferentiated markets (Fordism) and electronically mediated batch production for niche markets (post-Fordism). Within the industrial, organised, Fordist regime of production, there were dedicated assembly lines producing cars, washing machines, refrigerators, etc., and large numbers of unionised, relatively unskilled and well paid (mainly) male workers supported by housebound wives and a generous welfare state. In the 1970s, as a result of falling profits, saturated markets, labour unrest, production rigidities and massive oil price rises, this form of production came into crisis, and has been gradually supplanted by post-Fordism or a more flexible regime of accumulating capital. Here the assembly line is replaced by customised small-run production overseen by flexible, multi-skilled, non-unionised, individually contracted operatives and subcontracting firms, all of whom serve a highly differentiated market. For Linda McDowell, what also distinguishes this new regime of production is the role women play as casualised and sexualised members of the paid workforce in the rapidly growing service industries: these include retailing, personal service, tourism and recreation, and some sectors in the higher echelons of the service sector. She argues that post-Fordism has not only restructured production but also the sexual division of labour in workplaces and the division between home and paid work (McDowell 1991a).

The electronics which makes batch production and business services possible also allows the physical separation of production stages. Thus management, design, component manufacture, and assembly can be scattered across cities, regions and countries while also supporting an entirely new information technology industry and allowing the emergence of a new media-leisure-centred consumer. Beyond the workplace, those in employment can now consume more and more of these goods and services while, for the growing number of unemployed and part-time workers, the welfare state has been eroded and replaced by the idea and practice of user pays, and a rhetoric of individual initiative, independence and enterprise.

For geographers, it has been the socio-spatial consequences of such changes which have been of most concern. Thus the shift from a Fordist to a post-Fordist regime of accumulation has led to the deindustrialisation of cities, regions, even countries, and to the service or technology-based redevelopment of others; the latter include the sunbelts, silicon valleys and technopolises of California, Japan, south east England and Queensland. Globally, this has led to a differentiation of cities. On the one hand, there are those which house the international and regional headquarters of multinational manufacturing, trading and financial corporations as well as the many business and producer services needed by these firms—such as law, accountancy, advertising, freight, finance, research and development, real estate and insurance—for example, New York, London, Paris, Tokyo, Los Angeles, and Singapore. On the other hand, there are lower order cities which compete to support the regional production, national marketing or coordination activities of international firms; for instance, Vancouver, Sydney, Osaka, Munich, and Rome (Castells 1989; Sassen 1991, 1998; Zukin 1991). Within this global–regional economic dynamic there is another dimension much debated by sociologists concerning the extent to which globalising culture obliterates and homogenises local identities or produces and highlights difference (see Appadurai 1996; Featherstone 1996; Eade 1997).

In cities locked into this newly globalising economy there is greater social and spatial polarisation. The burgeoning class of technologically literate, skilled and well paid service workers occupy the desirable suburbs, inner city apartments and regenerated historic precincts. These exclusive enclaves are increasingly protected by elaborate measures to secure them from the growing numbers of unemployed manufacturing workers and marginal or part-time service workers who are often racialised and feminised and who live in decaying inner city areas or outer suburbs. It is these dual elements of glitzy, urban wealth and socio-physical poverty which typify the postmodern city (Fainstein, Gordon and Harloe 1993; Marcuse 1995).

These, then, are some of the major changes in the contemporary post-industrial space economy. A number of feminist geographers have been critical of this conceptualisation. For instance, Julie Graham highlights the ways in which such a narrative is totalising, exclusionary, 'stable, coherent and hegemonic' and has real political effects (1992, p. 393). In presenting 'a world in which capitalist development is the only road' (Graham 1992, p. 401), alternative conceptualisations and ways of envisaging exploitative social relations and sites of resistance, such as those within the household, are subsumed within a narrow notion of class-based social change. For Graham, along with Katherine Gibson, post-Fordism is a conservative rather than a liberating conceptualisation (Gibson-Graham 1996). If we recognise that production occurs not only in manufacturing plants but in the household, in the 'informal' sector, in the service sector and within the state, then, Graham concludes, the array of points for progressive political intervention multiply (Graham 1992).

The preoccupation of many postmodern geographers with the economic changes associated with postmodernism is both a strength and weakness, but it is also a different emphasis from that taken in other disciplines. In much of the postmodern writing emanating from literary and cultural studies as well as from some geographers, the focus on the text, the media image and the simulated or symbolic environment has directed attention away from production relations towards those of consumption (see, for example, Fiske, Hodge and Turner

1987). For some feminist geographers, such as Gibson-Graham (1996), Kay Anderson (1990, 1991), Robyn Dowling (1996) and Kathleen Mee (1994), the move to connect the cultural and social with the economic is a vital conceptual change; one which is impelled both by the shift in the global economy from production to consumption as much as by postmodern academic debates. As well as informing research on paid work (see chapter 2), a recognition of the interconnection between the socio-cultural and the economic has been most readily apparent in the consideration of particular environments within cities such as heritage precincts, gentrified neighbourhoods, the cultural festival, international hotels and shopping centres. Those environments involve built forms that are stylistically different from their modernist predecessors and encapsulate a range of postmodern socio-economic relations. It is from this particular aspect of postmodern style that other directions can be derived for the postmodern feminist geographer.

Postmodern Style

If the postmodern city is one of social polarisation, batch production and niche marketing, manufacturing decline and service sector expansion, it is also one where spectacle and commodification are increasingly important parts of the urban economy. David Harvey noted in the late 1980s a heightening of competition occurred between cities in the USA as their mayors and economic development agencies desperately competed against each other to secure hypermobile investment. City administrators searched for ways to generate and hold capital and also moved to create and mobilise new symbolic capitals through neighbourhood redevelopments, mass retailing, city festivals and spectacles (Harvey 1989b). These activities were often presented as indicating local distinctiveness—whether historical, ethnic or environmental—so that whole cities or their parts were marketed for easy consumption. The scale of such an exercise as well as the way in which it was done differed from the past. As Harvey noted:

> (T)he modernist penchant for monumentality—the communication of the permanence, authority, and power of the established capitalist order—has been challenged by an 'official' post-modernist style that explores the architecture of festival and spectacle, with its sense of the ephemeral, of display, and of transitory but participatory pleasure. The display of the commodity became a central part of the spectacle as crowds flock to gaze at them and each other in intimate and secure spaces like Baltimore's Harbour Place, Boston's Faneuil Hall, and a host of enclosed shopping malls that sprung up all over America. Even whole built environments became centrepieces of urban spectacle and display (Harvey 1989b, pp. 275–6).

As cities and precincts within them compete for capital investment, and as the urban environment itself becomes commodified, distinctions between high art and popular culture, between shopping and tourism, work and leisure, residing and visiting are eroded.

The international hotel—a place which accommodates the fleeting visits of the multinational investor and manager as well as the tourist—is one site which typifies much of this new economic and symbolic order. While the International School of modernist architecture produced a global proliferation of remarkably similar steel and glass office towers, high-rise housing blocks and hotels, the postmodern building, according to architectural historian Charles Jencks, is stylistically connected to its locality and thereby achieves a certain uniqueness. According to Jencks, this occurs through double coding, metaphor, humour and whimsy. The postmodern building is architecturally and socially a fusion of the modern with the vernacular, a process which can involve the local community in the design as well as in their use of the finished product (Jencks 1986) (See figure 4.1). For Short, such built forms are also heavily fortified by walls, gateways and security devices to produce an enclosed, privatised and separated space. This space can be either a building or an entire suburb ringed by gates, guards and walls to create a demarcated community, which is consciously separated from civic culture: what Short calls a postmodern bunker (Short 1996).

The Bonaventure Hotel in Los Angeles exemplifies a postmodern building (See figure 4.2). Built by the architect-developer John Portman—whose other works include various Hyatt Regency Hotels around the world—the Bonaventure is a place frequented by overseas and Californian tourists as well as by the local (white, middle-class) community. For geographers, cultural critics and architects who have written of this place, it both partakes of its environment and is separate from it, in its reflective glass facade and poorly demarcated doorways and blank walls, all above street level. Once inside, the space is vast: it is both enclosing, controlled and self-contained, confusing and disorienting, with a myriad of escalators symbolising and facilitating the constant movement in the hotel. For Frederic Jameson, these formal elements of glass, diagonals, mobility and steel as well as the experience of being in this space is postmodern, as it constitutes a 'new and historically original hyper crowd' (Jameson 1984, p. 81). The geographer Mike Davis, another observer of this hotel, sees it as representing a darker side of the postmodern condition. For Davis, who Excavat (es) the Future in Los Angeles, the intense social polarisation of this city, which is a consequence of its incorporation into the new global post-Fordist order, has necessitated the fortification of wealthy parts of the city as citadels for capital and conspicuous consumption. The Bonaventure's mirrored facade, elaborate security, obscure entrances

ANZ Bank, Queen Street

RMIT – Building 8

State Library of Victoria

RMIT – Storey Hall

Figure 4.1 Postmodern style, Melbourne 1999
Source: Louise Johnson

and self-contained opulent worlds, for Davis assert its physical and social distance from the mass of child, racialised and immigrant labour on which the economy of Los Angeles rests (Davis 1988, 1990). Surveying the same hotel with Jameson and

Davis, the geographer Edward Soja agrees with Davis's reading of its social and symbolic value: 'the Bonaventure has become a concentrated representation of the restructured spatiality of the late capitalist city: fragmented and fragmenting, homogeneous and homogenising, divertingly packaged yet curiously incomprehensible, seemingly open in presenting itself to view but constantly pressing to enclose, to compartmentalise, to circumscribe, to incarcerate...The Bonaventure both simulates the restructured landscape of Los Angeles and is simultaneously simulated by it' (Soja 1989, pp. 243–4).

Reading the city of Los Angeles and the Bonaventure Hotel encouraged Edward Soja to find new ways of describing and analysing what he saw. He therefore offered a number of traverses across, around, into and out of the city as he '(took) Los Angeles apart' (1986, 1996). This was one of various attempts to grasp the changes presented within such a city and a recognition that older ways of interrogating and theorising the city were no longer apt.

The Politics of Representation

Associated with the changing economic foundations of Western societies and their cities, there has been a fundamental change in the way people think about

Figure 4.2 The Bonaventure Hotel, Los Angeles
Source: Joe Hajdu

these places. This change derived from the experience of living in postmodern rather than modern cities, from reflecting on the technologies suffusing them (Baudrillard 1983; Lyotard 1984), from urban theorists' failure to understand them, and from devastating critiques of existing bodies of academic knowledge by socially marginalised and excluded groups.

The subsequent 'crisis of representation' has been prompted, in part, by feminist and postcolonial assessments of existing bodies of Western knowledge. These critiques have successfully exposed the exclusivity and partiality of previously hallowed universals. For, as Craig Owens notes, postmodernism is 'a crisis of cultural authority, specifically of the authority vested in Western European culture and its institutions' (Owens 1983, p. 57). The crisis of confidence in existing modes of representation has been created as much by the economic ascendancy and political independence of previously colonised and 'underdeveloped' Asian economies as by criticism in those countries questioning the intellectual hegemony of the West. Thus postmodernism can be seen as one part of postcolonial thought (see chapter 5). But, as Owens notes, women too, have long been excluded from dominant Western theories, histories and representations of the world.

In the 1970s, feminists such as Pateman and Gross (1986) recognised women as a marginalised group in Western thought; but in the 1980s, feminism itself was subjected to a postmodern–postcolonial critique for its focus on the experience of white, Western, heterosexual and middle-class women. This critique consisted of profound and angry evaluations by women who felt excluded from the grand visions of society presented by Marxism, Liberalism and Feminism. These views came especially from women in marginalised social positions in developed countries, such as blacks in North America, Asian and Afro-Caribbean peoples in England and Europe, non-English speaking migrants in Australia, and from colonised peoples in both the 'Third World' and in the West (Amos and Parmar 1984; Awatere 1984; Gunew and Yeatman 1993; hooks 1982, 1990. See chapter 5). In this reassessment, not only was feminism revealed as partial but the whole of Western thought—founded on the Enlightenment faith in reason, truth, progress, the unity of individuals and the separation of people from their environments—was increasingly thrown into question. This critique produced a sense of knowledge as relative, provisional and grounded rather than absolute, definite and universal, and the world as knowable not through sense perception but through words, stories or discourses uttered by those in various positions of power.

The task of exposing the exclusion of women, racial and ethnic minorities in existing bodies of academic and social knowledge went well beyond a liberal feminist noting of absences; it involved using a key tool of postmodern cultural criticism, that of textual deconstruction. This approach derives from the histor-

Figure 4.3 Postmodernism: A feminist view
Source: Judy Horacek, 1997

ical research and theorising of Michel Foucault. In enquiring into the minutiae of everyday life in nineteenth-century France, Foucault demonstrated that identities are constituted from the discourses and power relations in which people are situated and engaged; and that this was especially so with those discourses emanating from various professions and arms of the state. By discourses, Foucault meant statements, whether verbal, written, architectural or visual, made by socially influential groups or individuals. Such an origin gives these utterances a particular currency or authority which is then actively engaged with by others. A discourse analysis involves unpacking the nature of any message, locating it socially, and engaging with the way in which others both receive and reject it. Foucault's studies of institutional practices and regulations, and of the discourses which define and construct sexualities, the prisoner and the mental patient revealed that people were neither heroic individuals (nor fundamentally gendered) but historically and discursively constituted subjects and bodies (Foucault 1972, 1973, 1977, 1979).

In discussing the French penal system for example, Foucault set out the ways in which the laws, courts and regulatory practices defined what it was to be criminal. The architecture of the prison enforced certain ways of constituting, seeing and policing prisoners. Thus Jeremy Bentham's invention, the panopticon—a circular viewing tower which allows uninterrupted views over a number of radiating cells and corridors—facilitated the gaze of authorities over the prisoner. Along with rules and procedures devised by the police, prison authorities and contained within the legal code, which comprised a set of discourses created and enforced by those in particular positions of power, the built form of the panopticon actively assisted in the ongoing construction and policing of the criminal body. By examining other discourses, such as those surrounding the definition of the mentally ill and the nature and meaning of sexuality, Foucault was able to construct histories of mental patients and sexuality, and also to offer a technique for the analysis of discourse at other sites and in other environments.

Foucault's insights into the surveillance practices of the panopticon have been directly incorporated into designs for, and analyses of, the postmodern city. Here is a description of a shopping centre in Los Angeles:

> The King center site is surrounded by an eight-foot-high, wrought-iron fence comparable to security fences found at the perimeter of private estates...Video cameras equipped with motion detectors are positioned near entrances and throughout the shopping center. The entire center, including parking lots, can be bathed in bright...lighting at the flip of a switch. There are six entrances to the center: three entry points for autos, two service gates, and one pedestrian walkway...both service gates remain closed and under constant closed-circuit video surveillance, equipped for two-way voice communications, and operated for deliveries by remote control from a security 'observatory'. Infra-red beams at the bases of light

fixtures detect intruders who might circumvent video cameras by climbing over the wall. The 'unobtrusive' panopticon observatory is both eye and brain of this complex security system...It contains the headquarters of the shopping center manager, a substation of the LAPD, and a dispatch oper-ator who monitors the video and audio systems as well as maintaining communications 'with other secure shopping centers tied into the system and with the police and fire departments' (Davis 1990, p. 243).

While Elizabeth Grosz and Judith Butler agree that Foucault's notion of the body constituted in and through various discourses and spaces is vital, they suggest that it needs acknowledging that gender is central to that process (Grosz 1993, 1994; Butler 1990, 1993). Thus, in extending Foucault's discourse analysis, and by adding the technique of textual deconstruction, some feminist geographers have begun to examine the constitution and regulation of women's bodies in space; for example, in workplaces (see chapter 3), in gymnasiums (Johnston 1995), in sport and in shopping centres (Longhurst 1994, 1995) (see also Moi 1985; Weedon 1987).

Deconstruction is a way of reading and analysing texts or discourses which exposes the systems of power which inhere in them. These systems of power authorise certain representations while blocking, prohibiting or invalidating others. Arising from a particular view of human development and the role of language within it, deconstructive practice has a long history in literary studies and psychoanalysis. Many feminists were deeply troubled by the patriarchal undertones and assumptions in Sigmund Freud's work until Juliet Mitchell suggested that his work was an analysis of a patriarchal society not a celebration or endorsement of it. Feminists still have a contradictory and tense relation to psychoanalysis—some using it comfortably and others remaining hostile. Mitchell, however, pointed out that Freudian psychoanalysis uncovered the ways in which male anatomy and sexuality were privileged in the course of childhood development while female desire and identity were constructed in terms of lack, an absence and set of desires constantly striving to be fulfilled (Mitchell 1979). The analysts Jacques Laçan and Jacques Derrida who combined psychoanalyis with structuralist linguistics, see the moment when an infant starts to acquire language as fundamental to its enculturation. They argue that language is consti-tutive of particular worlds; hence the system of language which an infant enters is both critical to its engagement with a particular world which is profoundly gendered and endlessly unstable. Words, as utterances signifying or representing something, do not have fixed meanings but rather derive their sense from the play of difference within wider fields of meanings. This intertextuality, the broad field of possible meanings, is open to multiple and ever shifting interpretations

(Sayer 1993). What any word signifies or means therefore is deferred, it can never be fixed but is endlessly unstable. For any utterance or written word, there is a range of possible meanings, which are both present and absent in any statement. A deconstructive reading becomes the means of articulating these inclusions and exclusions, of marking out the statement's unstable meanings, and the power dynamics associated with the process of creating, circulating and interpreting them (Bordo 1993; Weedon 1987).

Figure 4.4 The waiting room of postmodern meaning
Source: Judy Horacek, 1997

In this schema, language is unstable yet also critical to the creation of iden-tity. Hence identity is open and contestable. For postmodernists, the idea of a fixed, clear and unambiguous individual, with a definite and unchanging notion of self, is a fiction. Rather a person is gendered from the moment they enter language, and their identity is a shifting, contested and fluid amalgam of the various discourses with which they actively engage. For women, this view, combined with the Freudian notion of female desire, raises the possibility of 'woman' as an unstable category within which an array of identities can be defined and cohere around dimensions of sexuality, race, ethnicity, etc. Furthermore, the fixing of any one meaning is a political act. Thus, as Judith

Butler argues, identity categories are never merely descriptive but always normative and hence exclusionary. Thus it becomes a necessary political and textual task to trace the ways in which women are included and excluded within discourses, to discover who effects that inclusion and exclusion, and to delimit the ways in which women are defined in particular ways rather than in others (Butler 1990). The notion of what it means to be a 'woman' is thus permanently open and available for women to redefine, even though female identity and desire is also the subject of many confining discourses; especially those to do with health, beauty, fashion and food. It is this woman who experiences and engages with the postmodern city. As Elizabeth Wilson writes: 'In postmodernism the city becomes a labyrinth or a dream. Its chaos and senselessness mirror a loss of meaning in the world. At the same time, there may be an excess of meaning: the city becomes a split screen flickering with competing beliefs, cultures, and 'stories'. This play of unnerving contrasts is the essence of the 'postmodern' experience...Everything is the same and nothing is quite real' (Wilson 1991, p. 136).

In attempting to apprehend this place, Wilson writes in an evocative, poetic, nonlinear and unstructured style. For her, the contemporary city is a place of contradictions: of sameness and difference, excitement and fear, pleasure and danger, mediated and fragmented by media and computer technologies. In this place and space, individuals become decentred and disoriented yet potentially free to explore the many possibilities now open to them. This is especially the case for women who suffer the limitations of a patriarchal city yet who can also seize the many new challenges for autonomy which it offers (Wilson 1989, 1991). In Wilson's description, the idea of a fixed, anonymous individual buffeted by structural forces is replaced by a gendered and sexualised woman who is endlessly open to new desires, identity formations and makeovers. This woman is also located in time and space so that any theorising or representation of her must acknowledge both her fluidity and specificity. This view of women derives from feminism as much as from postmodernism. But the deconstruction of a fixed identity, which is derived from an emancipatory politics, is part of the postmodern move to reimagine the individual as the fragmented subject and product of discourse.

Thus a fundamental element of postmodernism has been a crisis in old forms of representation. Former assumptions about truth, an all embracing theory, universalism, empirical knowledge and the individual have been shown to be partial and limited conceptualisations which are as much a product of the Western imperial mind as of the patriarchal mind. These assumptions are replaced by a view of knowledge as produced by discourse; this knowledge is relative, grounded and the product of subjects who occupy certain positions of power according to their gender, class, race and ethnicity. Thus the linguistic turn in postmodern thought and the technique of textual deconstruction have offered a powerful set of tools to feminist geographers; while rethinking the world as a series of texts—a rethinking which includes buildings and the sexed bodies that move in them—offers new directions for postmodern feminist geographies. In the next section, I briefly consider postmodern feminist geography before using aspects of this framework to examine a regional shopping centre in Sydney.

POSTMODERN FEMINIST GEOGRAPHIES

As with other feminists, some feminist geographers have actively engaged with postmodernist thought while others, such as Elizabeth Bondi and Mona Domosh, see the dominance of men, and the focus in postmodernist work as having little to offer feminist geographers (Bondi 1990, 1992; Bondi and Domosh 1992). This problem, however, has led to some of the most incisive work in feminist geography. One of its vital tasks has been critiquing postmodern geographers for ignoring feminist insights, and for failing to see women as an oppressed, differentiated and politically active group. Feminist deconstructive techniques have been instrumental in this critique.

These deconstructive principles have been used in examining the work of those who define postmodern geography. Two important writers here are Edward Soja and David Harvey, whose work has been influential well beyond the discipline. Harvey's *Condition of Postmodernity* (1989a) and Soja's *Postmodern Geographies* (1989) have been the subject of critiques from feminist geographers such as Rosalind Deutsche (1991), Doreen Massey (1991), Liz Bondi and Mona Domosh (1992), Steve Pile and Gillian Rose (1992). This criticism asserts that Harvey and Soja's books:

- contain an unexamined sexism and phallocentrism: the authors fail to recognise their position of discursive power as white, heterosexual, male geographers
- contain a construction of the postmodern condition which completely ignores women and feminism
- do not acknowledge that patriarchy and racism are a critical part of social relations and central forms of oppression, in both modernism and postmodernism, nor that feminist, anti-racist and postcolonial struggles are significant political movements
- fail to see themselves as offering totalising, objective theory rather than as engaging with a postmodern deconstruction of grand narratives which acknowledges diversity and difference.

These critiques have exposed the sexism in these two texts in excluding women and feminism; they have also detailed the ways in which Harvey and Soja's work has mobilised various textual strategies to place them, as male authors, in an unassailable and unreflective position. Thus feminist deconstruction has revealed the phallocentrism in these texts. For example, Rosalind Deutsche notes how David Harvey constructs himself as a universalising masculine subject who 'discovers' rather than constructs a certain postmodern reality. Harvey thereby claims universal knowledge while actually occupying a very partial and privileged position (Deutsche 1991). Doreen Massey also unpacks the ways in which Harvey's book uses a range of female images and that, while he does not comment on them, he claims they are somehow representative of women; she also suggests that he does this in a way which affirms an unproblematised masculine point of view and a negative view of women (1991).

Deconstructive techniques have also been central in Gillian Rose's *Feminism and Geography* (1993). In one of the few book-length feminist geographies, Rose provides a critical reading of mainstream geography texts as well as a number of feminist geographies. Her reading uncovers the ways in which 'time geography' and socialist feminist geography, for example, mobilise and assume certain binary opposites such as nature–culture, body–mind, emotion–reason, home–work, and reproduction–production to create a body of geographical thought and practice. Rose argues that though this work is progressive in many ways, it also rests on these conservative dualisms. The nature/culture divide has long been associated with the female/male distinction in which women are usually negatively identified with nature and with bodies, emotion, home and reproduction. Rose's critique is powerful: it highlights the way in which women are variously inscribed in geographical works, the utility of deconstructive practice and how other dimensions of difference—such as race and ethnicity—may be acknowledged but are rarely given a central place (see chapter 5).

In addition to using deconstructive techniques to interrogate the discipline, feminist geographers have also been concerned with gendered bodies in space. Drawing on the work of Judith Butler and Elizabeth Grosz, feminist geographers have begun to recognise that women and men move through spaces in quite different ways and that these differences are related to their own physical form or corporeality. Furthermore, accepting Grosz's arguments which draw on Foucault, that the body carries traces and inscriptions of a range of medical, legal and advertising discourses, feminist geographers have studied how women's bodies move in space, how they move in response to ideals and discourses, and how they are made into particular forms through sport and weight training (Johnson 1989). In New Zealand, Robyn Longhurst's study of the pregnant body in sport, and in a city shopping centre challenges the mind–body distinction and captures the ways in which this body is discursively constructed as unwell, frail, disorderly and transgressive (Longhurst 1994, 1995). Lynda Johnston has also studied the gymnasium as a space in which women both remake their bodies, thereby challenging various stereotypes of the weak effeminate and frail female body, and affirm other stereotypes created and sustained by the body-building industry (Johnston 1995). The scope for work on the sexed body is vast (see Johnson 1989), and I have taken this concern, along with insights derived from postmodern geographies—such as post-Fordism and textual deconstruction—into my study of a quintessentially postmodern space. This is Westfield Shoppingtown, a regional shopping centre in Parramatta, western Sydney.

A POSTMODERN REGIONAL SHOPPING CENTRE

The regional shopping centre in its initial manifestation was one of a number of building forms which included the freeway, the suburban factory, the international airport and the high-rise office building, which characterised the international post-1945 style. With its standardised layout, identical shops, greenfield location and orientation towards suburban families and the car, the regional shopping centre typified US modernity. Recently, the regional shopping centre has changed: it has become one of many places which represent a key element of the post-Fordist economy, where postmodern style is expressed and identities negotiated. The contemporary mall is no longer located in a sea of undifferentiated suburbia around a single city centre but is part of a dispersed series on the edges of cities. Retailing has become a core rather than a peripheral economic sector as it is increasingly connected to other areas of a booming service economy—personal and business services, recreation and tourism—while the actual experience of shopping is linked to the formation of identity. Thus regional shopping centres, like those who occupy them, are undergoing makeovers—both physical and conceptual—as they assume the mantle of postmodernity. Before examining one of these centres in Australia, the centre needs locating in the larger socio-economic context of post-Fordist transformation. In the case of Westfield in Sydney's western suburbs, this centre is situated in an increasingly globalised city and in an urban economy with a booming service sector, social and spatial polarisation, ethnic diversity and densification. A postmodern feminist geographic analysis involves deconstructing the centre as a text, and seeing it as a place in which gendered bodies and identities are forged, challenged and made anew.

Sydney as a Postmodern, Global City

In 1966, the British geographer Peter Hall placed Sydney near the top of a group of 'sub-global' cities, which included Singapore, Hong Kong and Los Angeles. Heading the hierarchy were the World Cities of New York, London and Tokyo. Inclusion in the hierarchy required cities to have major business centres which were also a focus of national and international government and centres for trade, finance and communication (Hall 1984). Twenty-five years later, Saskia Sassen (1991, 1994) joined John Friedman (1986) and Anthony King (1989) in distilling global economic status and influence as the key elements which distinguished global cities. Again, Sydney was included, as one of the key second order cities which are integral to the operation of multinational finance capital in the Asia–Pacific Region.

In his discussion paper for the New South Wales Department of Urban Affairs and Planning, 'Sydney as a Global City', the geographer and planner Glen Searle showed how Sydney connected Australia to the global economy (Searle 1996). Searle identified this link as a function of the city having the nation's main international air and communications gateway, its national dominance as a port for export-related businesses, and its place as Australia's major financial centre and thus the focal point for most international finance transactions. In 1994, the city's Futures Exchange was the eleventh largest international exchange, and included commodities, currency and share futures. So too with banking; in 1996 Sydney had 37 of the 52 head offices of banks with a full Australian licence, while in 1988, 155 of the 185 Australian head offices of foreign banks were located in Sydney. The concentration of national and international banking activity makes the city a centre for multinational capital command and control functions. This takes two forms: first, Australian companies with overseas interests. Thus of Australia's top 100 companies in 1989, 60 had headquarters in Sydney compared with 29 in Melbourne; this compares with 1984 when only 45 of the top 100 Australian companies were based in Sydney and 41 in Melbourne. Second, the overseas-based multinationals who are also locating their regional offices in Sydney. Thus by the late 1980s, about 150 international institutions had their head office in Sydney compared with 43 in Melbourne; 39% of the regional headquarters of the top 20 transnational corporations in each of the fields of accounting, advertising, management, consulting and international real estate in 1990 were in Sydney, compared with 32% in Hong Kong, 13% in Tokyo and 6% in Singapore, other Australian cities sharing the rest.

In addition to these developments, Sydney is the main destination for foreign investment in Australia. In 1994–95 it absorbed 42% of the total foreign investment in Australia; in tourism and real estate New South Wales had nearly 50% of the national total; of the $5 billion of real estate investment in Australia, 46% was in New South Wales, mostly in Sydney, so that 15% of city office properties were foreign-owned. While such investments have produced skyrocketing property prices in Sydney, these costs relate not to other Australian cities but to similar cities in the global economy. Compared with other major cities in the Asia–Pacific Region, Sydney has lower costs and charges for real estate purchase and rental and infrastructure services. For example, in March 1996, the purchase price per square metre of prime CBD office space in Sydney was at least 75% below that in Tokyo, Hong Kong and Singapore and significantly less than that in Taipei, Beijing and Seoul. The basis of comparison is not with other Australian cities but with cities at comparable points in the global hierarchy of cities (Searle 1996).

As a consequence of the concentration of multinational corporate offices, communications, tourism, business services and trade, Sydney's employment structure increasingly mirrors that of other World Cities in the growing number and proportion of people in the finance and business services sector. At 17% the proportion is now approaching that in New York and London in the mid-1980s and is significantly ahead of Melbourne's 12.7% (see table 4.1).

	1991		1996	
	No.	%	No.	%
Retail trade	191493	12	213291	13
Manufacturing	213662	14	214753	13
Property and business services	151407	9.7	211441	12.6
Health and community services	125663	8	151844	9
Other	874389	56	884132	53
Total	1556614	99.7	1675461	100.6

Table 4.1 Employment by selected sectors, Sydney 1991 and 1996
Source: ABS catalogue no. 2017.1, 1996

Most of the population increase of the city now comes from net overseas migration. Thus in 1991, 1.1 million or 30% of the city's population were born overseas, 21% from a non-English speaking country and 5.3% from the main east Asian countries. Sydney's ethnic diversity and high levels of overseas migration means that it has significant numbers of residents who are proficient in both the main east-Asian languages and in English, the global commercial language. High levels of immigration have also added to the social differentiation of the city, so that new arrivals often move into the inner and outer western suburbs of the city. Migrants with little capital are excluded from the city's property market and thereby exacerbate the patterns of social polarisation which have come to typify the city (Murphy and Watson 1997).

There are a number of things which make Sydney a significant global city. They include its connections with international finance, its role in the command and control functions of major multinational corporations, its employment structure, its urban land market, its ethnic make-up and position as the national and regional centre for trade, and its communications and business sector. In addition to this economic dominance, its array of institutions and products also indicate that Sydney is Australia's cultural capital: it is the centre of the nation's advertising industry, major arts funding bodies (the Australia Council) and arts companies (the Australian Opera), the Australian Broadcasting Commission, head offices of television and newspapers and of the Australian film industry. In structural terms, Sydney is thus Australia's pre-eminent post-industrial global city. As Sassen observed: 'the developments of the 1980s—increased internationalisation, a strong shift toward finance, real estate and producer services—contributed to a greater concentration of major economic activities and actors in Sydney. This included a loss of share of such activities and actors by Melbourne, long the centre of commercial activity and wealth in Australia' (Sassen 1998, p. 210).

Alongside the increasing global connections of Sydney, the growth in its high level service functions and booming real estate market, its population has increasingly bifurcated along class and ethnic lines. Across the city, there are marked divides between the affluent eastern and northern suburbs and the west of the city; this divide is registered in unemployment rates, income levels, educational levels, public housing, proportion of single-parent families, newly arrived immigrants and service sector job growth and employment.

While such patterns have long existed in Sydney, they have been accentuated by the current round of restructuring and globalisation of the city (see chapter 1 for the experience of other Australian cities). As Murphy and Watson note: 'This dichotomy has become more marked since the mid-1970s with much of the increase in unemployment being in areas where there are more unemployed than the regional average. Areas of high unemployment are also areas where the workforce is younger, less well educated, more likely to be born overseas and more likely to work, or to have worked, in factories' (1997, p. 97).

It is this economic and spatial context which contain Sydney's regional shopping centres.

Westfield Parramatta

Located in an inner western suburb, the extended market areas served by Westfield Parramatta straddle two economic and social sectors of the city: it is located squarely in the 'deprived west' and adjacent to the northwestern hills district. While the notion of a deprived west is a contested and exaggerated term, this is still a region which has been hardest hit by the collapse of manufacturing employment, and which has witnessed growing displacement from home owner-ship and accommodated a large proportion of low income and newly arrived non-English speaking migrants (Powell 1993; Grace et al. 1997). Westfield Parramatta is also accessible from the more affluent northern suburbs. Thus as table 4.2 indicates, the immediate area served by this centre is one of relatively low incomes (averaging $16 000 compared to the Sydney average of $18 500), high levels of overseas born (41% compared to the overall Sydney level of 32%), and with a high proportion of renters rather than home owners (48% compared to the Sydney average of 32%). Extending north as well as west, the broader market area approaches the Sydney socio-demographic average.

Key socio-economic characteristics	Primary trade area	Secondary trade area	Sydney
Population density (per sq km)	2 440	2 008	308
Average per capita income ($)	16 287	15 881	18 485
Average household size	2.7	3.02	2.82
Population change 1996–99 (%)	1.0	0.0	1.0
Housing status:			
Owner/purchaser (%)	51.9	72.4	68.2
Renter	48.1	27	31.8
Birthplace:			
Australia (%)	58.6	63.5	67.8
Overseas (%)	41.4	36.5	32.2

Table 4.2 Demographics in the Westfield Parramatta and Sydney areas, 1998
Source: *Westfield Market areas*, 1999

The centre is also located in a business centre with a range of edge city func-tions and employment. As the 1998 company report noted, the Westfield Parramatta and Chatswood complexes on Sydney's north shore are the only two regional shopping centres which serve both a suburban and a central business district. Their market areas and location, along with substantial investment in expansion and promotion, has led to spectacular increases in turnover and prof-itability. Thus in 1995 Westfield Parramatta's turnover was $300 million, but after an expansion and refurbishment this increased to $550 million in 1998. By 1998, Westfield Parramatta was the largest shopping centre in the country, with 389 shops covering 128 774 square metres across five levels, and valued at $350 million (Westfield Holdings 1998). While vast by Australian standards, Westfield Parramatta remains far smaller than some of the North American mega-malls such as West Edmonton Mall which has 380 000 square metres and

The Mall of America with 400 000 square metres in Minneapolis (Goss 1992; Shields 1992).

Initially constructed in 1974 as a relatively small, integrated shopping mall at the southern end of the main retail street in Parramatta, namely Church Street, the centre was physically linked to its locality in a host of ways. Adjacent to a railway station, it has become the focal point for a maze of bus routes and encircled by a set of major roadways. However, as a covered-in regional centre located away from the main street retail area, Westfield inevitably competed with existing retailers and, as with many other suburban centres, ultimately contributed to the demise of the strip retailing precinct around it. But this event was not uniform or uncontested. As Hawkins and Gibson show, the mid-1990s expansion plans of Westfield at Parramatta challenged not just small scale retailers and their local government representatives but also other large scale investors; in the case of Parramatta, Merlin Investments who wanted to shift the retail focus of the locality away from Westfield and the railway towards the river and the old retail core of the city (Hawkins and Gibson 1994). Merlin's plan for redeveloping the southern end of Church Street was ultimately defeated by their financial woes, court action and the power of Westfield. But the battle for retail hegemony in Parramatta has left some significant scars on the landscape—in the form of a moribund precinct at the northern, river end of Church Street—but it has also left an enlivened city mall in which community activity, restaurants and retailing thrive during the week (see figure 4.5).

Westfield Parramatta was therefore never on a greenfield site; it was strategically located close to a range of public transport routes and a densely settled, ethnically and economically diverse catchment (see table 4.2). As a centre it has capitalised on the rising densification and changing household structure of Sydney's population. Thus as Trevor Sykes noted in *Australian Business Monthly*:

> Two trends have been driving Australians towards regional shopping centres. One...has been urban consolidation, which means our suburbs are becoming more densely occupied. Another is that there are fewer people living in each house of flat...As each house still needs a refrigerator, television and some furniture, demand for white and brown goods has been good. Surprisingly, Westfield's research also shows that demand for food has increased (Sykes 1993, p. 40).

The centre has also become more densely occupied: spread over five levels with a sense of activity rather than space. Westfield Parramatta is one of 38 owned by Westfield Holdings across Australia. It nests in a set of family trusts

Figure 4.5 Church Street Parramatta 1999: Retail defeat and community revival
Source: Louise Johnson

and holding companies overseen by the Lowry family which now extend into the USA, New Zealand and Malaysia (Clafton 1995). The company first ventured off shore into the US market in the late 1970s, and by 1993 either owned or was managing seven shopping centres. In 1997 the company took over TrizeHahn in the US thereby acquiring a string of centres which, in 1998, totalled thirty-eight. These acquisitions have led to a focus on a set of markets and regions—what Westfield describes as a 'clustering' and 'multiple market presence' across six major markets: those of St Louis, Washington DC, Connecticut, San Diego, Los Angeles and Northern California which has the main concentration of nineteen stores. Combined with the purchase of a 47% stake in St Lukes in New Zealand—delivering eight wholly owned and two 50% owned shopping centres in Auckland (7), Wellington (1), Hamilton (1) and Christchurch (1)—and a 10% share in the Suria Kuala Lumpur City Centre project in Malaysia, these activities make Westfield a major player in both Australian and global retailing (Westfield Holdings 1998).

Westfield Holdings is therefore a large scale multi-national shopping centre owner and manager. Through its shopping centres the company oversees and shapes a highly centralised but also dispersed retail sector which is assuming greater significance in the Australian economy. Table 2.1 showed the major decline in manufacturing employment which dropped from the 1950s figure of 28% of the workforce to 16% in the 1990s, and the rise in service sector employment from 20% to 34% in that time, with this occurring mainly in Financial and Business Services and also in Tourism and Recreation. It also showed that the 4% fall in the proportion employed in 'Wholesale and Retail', which increased from 16% in 1955 to 18% in 1975 and thence declined to 14.8% in 1995 (Fagan and Webber 1999, p. 83), occurred despite the significant increase in the numbers employed in this sector. The relative fall in retailing does not indicate a weakening of this sector. On the contrary, retailing remains a significant economic sector; profitability is high for the larger scale operators, while shopping is increasingly linked to the other growth sectors of the economy such as recreation, tourism and business services which are increasingly a part of retail precincts.

The retail industry comprises businesses primarily engaged in the resale of new or used goods to final consumers for their personal or household consumption. Retailing is the third largest of Australia's industries; in terms of sales of goods and services it rates behind manufacturing and wholesaling. In 1991–92 retail outlets recorded a turnover of $114.3 billion (ABS Year Book 1995). While dominated numerically by small businesses, with 96% of businesses employing fewer than 20 people, small businesses accounted for only 44% of the industry's turnover during 1991–92. In other words, while only a small number of retail businesses employ large numbers of people, these are the ones which generate most of the turnover. Thus in 1993 the four largest retailers in Australia accounted for 21.5% of sales (ABS Year Book 1995). In the clothing trade, levels of concentration are even higher with sixty-seven cents in every dollar spent going to Coles Myer through one of its many companies and brands. Retailing is therefore a mix of some large scale corporations and a huge collection of smaller, often family-owned independent satellite, franchised and subcontracting operations. In this structure retailing assumes many of the distinguishing characteristics of the post-Fordist economy.

The industry's workforce is dominated by a highly gender segregated and increasingly part-time workforce (see table 4.3). Not only is this sector one of the largest employers of women but wages are low, the work is 'unskilled' and increasingly casualised and non-unionised. The association of certain products with masculinity and femininity, a key element in their promotion and purchase, extends to who should actually handle and sell the product or deal with the routine tasks of taking fast food orders and handling cash at registers. This can be part of the sexualisation of the commodity, whether a car, meat, a meal or an item of women's or men's clothing. It is women who primarily handle and sell food and clothing but men who process and sell meat, fish and poultry and cars. The gendering of commodities is related to the sexualisation of consumption and the use of desire in retailing environments. How these regional shopping centres are constructed and how such desires are mobilised are other elements highlighted by a postmodern feminist geography.

Retail activity	Males		Females	
	Full-time	Part-time	Full-time	Part-time
Supermarket /grocery	43591	44298	43880	94289
Specialised food	76752	61210	42938	111258
Department stores	9950	14377	23376	50879
Clothing and soft goods	14767	8350	33739	39977
Furniture, houseware appliances	59309	7659	24120	17930
Recreational goods	21014	8536	13502	19434
Household equipment repair	17723	4185	2485	3551
Motor vehicle retail	47052	674	9030	2954
Motor vehicle service	139813	14801	16984	26740
Total retail	473632	174496	254704	420274

Table 4.3 Male and female employment patterns for selected retail groups, Australia 1999
Source: ABS data supplied August 1999

Westfield Parramatta—A Postmodern Space?

In addition to capitalising on the complex pattern of social polarisation across this city, Westfield Parramatta represents other elements of the post-Fordist economy, in the structure of the retailers who rent space, and the shoppers who move in it. It is a vast retailing space owned and controlled by a major multinational corporation in which there is a mix of large scale retailers—including the major department store anchors of Grace Brothers, David Jones, K-Mart, Target and Woolworths—as well as other major-name retailers of food, toys, sporting goods and soft furnishings. In addition, this space accommodates a large number of smaller outlets (many of which are also national and international franchises) in which many thousands of part-time female workers labour, and through which some 343 000 shoppers circulate each week, making an extraordinary total of 18.6 million shopping visits a year (Westfield Annual Report 1998).

Can the multi-level space of Westfield Parramatta be read as a postmodern text or as expressing a postmodern aesthetic? Is it a modern or a postmodern space?

In her analysis of contemporary Australian shopping centres, the cultural critic Meaghan Morris identified a number of critical architectural elements: these were their monumental presence, familiar layouts with variations which derived from particular management decisions and localities, a unique sense of place (or a myth of identity), and spaces which are fluid and indeterminate. For Morris, the question of locality is just how does a shopping centre reflect its socio-economic or physical location. Morris concludes that postmodern shopping centres are overwhelmingly and constitutively paradoxical (Morris 1988).

In its five floors towering across four city blocks, Westfield Parramatta conforms to the first of these elements. Whether its layout is 'familiar' in the sense of being tied to its locality and negotiable by outsiders must await surveys of shoppers, though the layout of regional shopping centres in Australia is reputedly remarkably similar. The basic form in Westfield Parramatta of anchor department stores—one upmarket and the other a discounter—linked by long corridors full of specialty stores, the ever-present multicultural food court and the vertical separation of entertainment, food and clothing gives this centre a similar design to others. Some suggest that familiarity with one centre means others are comprehensible (see Beddington 1991; Gruen 1973). However, when I first entered Westfield in May 1999, I was struck by how incomprehensible it was compared to another greenfield regional shopping centre I had studied, that of Highpoint in Melbourne (see Johnson 1996b). Westfield had two rather than one major anchor department stores, and they were placed, as figure 4.6 shows, not at either end of a standard dumbbell shaped complex connected by long corridors flanked by specialty shops, but adjacent to each other. Furthermore, the differentiation by level—between entertainment on Level 1, fresh food on Level 2, low order retailing on Level 3 and higher order (in the sense of being more stylish, exclusive and expensive as well as less frequently purchased) on Level 4—was nowhere near as clear or as consistent as the standard dumbbell complex; a very different pattern was apparent. There was no exclusive precinct but rather up-market—but not designer—shops were scattered across the space interspersed with a range of other clothing, homemaking and apparel shops. This pattern conformed more to the configuration Fiske, Hodge and Turner noted in 1987 for Brisbane's Indooroopilly Shopping Centre; they suggest this pattern is an expression of the low socio-economic status of the market area when compared with Sydney's Centrepoint (Fiske et al 1987).

This is one element in the incomprehensibility of this place. The confusion also derived from there being five floors of shops, and many multi-level access points from five car parks and five streets. Westfield Parramatta, like the other CBD-style centre in Chatswood, was therefore vertically configured rather than laid out horizontally as are more familiar greenfield centres. This centre's apparent lack of pattern contained another order which meant that it differed from other modern centres: its main focal points were located not near the anchor department stores, nor near retail shops but near food and leisure outlets. These dual elements of fresh, semi-processed, take-away and restaurant/cafe-style food outlets and leisure-oriented shops (selling sporting, photographic, beauty, games, films and branded merchandise) are not on separate floors but integrated into the core flows of the centre (see figure 4.6). Thus the major additions to the complex since 1995 have been expanded food courts, more restaurants and cafes, additional spaces given over to fresh food retailing and to specialist leisure and promotional retailers, namely INTENCITY and Disney (Annual Report 1998). These new outlets, which are often located close to the major car park entrances, are now the critical hubs around which circulation patterns revolve. A deliberate strategy by management, the shift from retailing towards food, leisure and entertainment has meant profound differences in the ways in which this centre was laid out and functions. As the 1998 Annual Report on redevelopment plans for the whole Westfield group observed: 'An important element in these developments has, and will continue to be, a focus on lifestyle and entertainment which has proved so successful in attracting new shoppers and ensuring their visits are not only longer, but more enjoyable' (Annual Report 1998). The next section takes up the significance of this point in the creation and modification of identities.

The question of whether such a configuration and centre generates a unique sense of place is difficult but vital. The history and location of the centre is certainly placebound, but its architectural and retailing elements are only partially connected to the socio-economic profile of its trading areas. Once in the centre, with its huge department stores, long shopping corridors, multi-level balconies, food courts, towering atrium and theatre complex, you could be at

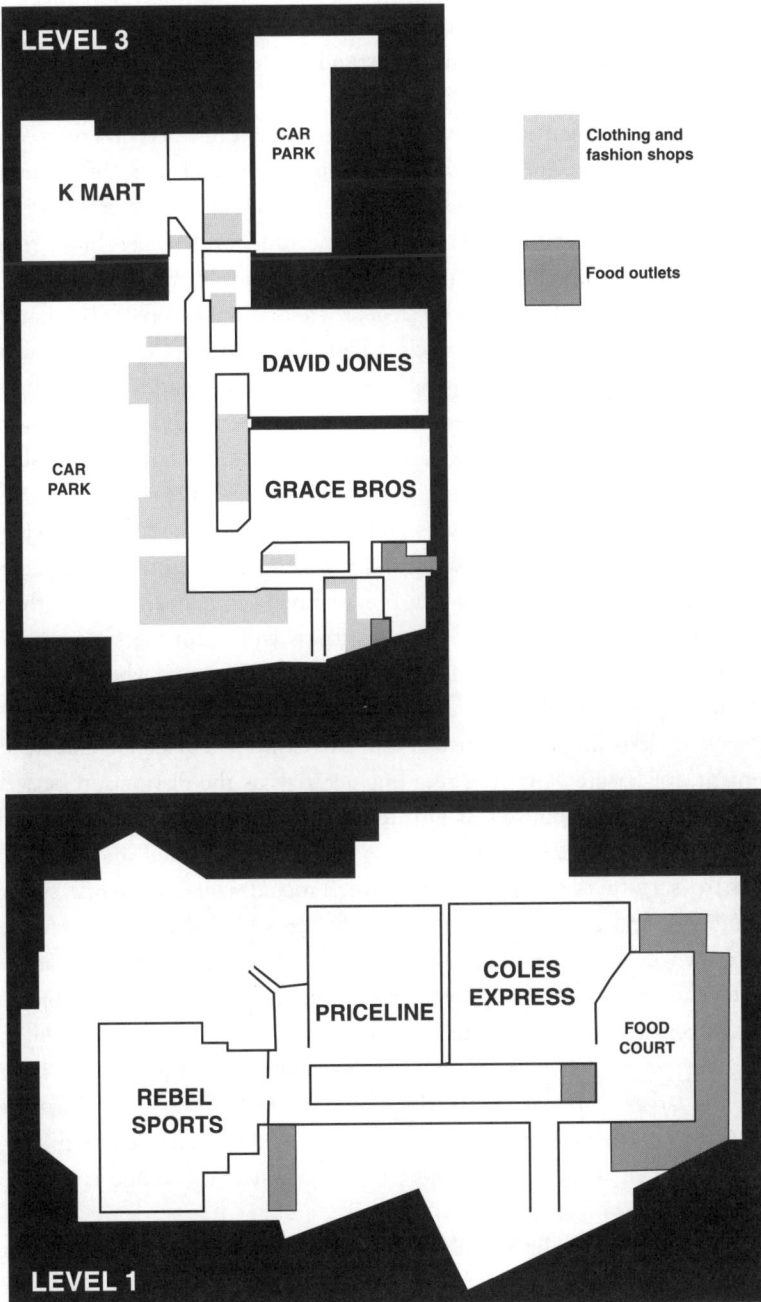

Figure 4.6 Artist's impression of levels 1 and 3 floor plans, Westfield Shoppingtown, Parramatta
Stephanie Thompson

any centre of similar size across Australia (or even the Western world). These elements of geographical specificity and standard design echoes Jencks discussion of double coding (Jencks 1986).

There are other elements of postmodern space present here. As observers of the postmodern have noted, in the extended trading hours and range of services—from retailing through to theatres, financial services and electronic games—places like Westfield Parramatta challenge the distinction between high art and popular culture, retailing and leisure/recreation/tourism. In the towering atrium complete with potted palms and open stage, there is a simulation of elsewhere, of other places, an ongoing urban spectacle, a pastiche of various elements. Here the hypercrowd is made: that passing parade of shoppers who see this layout and set of spaces as usual, normal, self-referential and comprehensible, even though it is completely artificial. Importantly too, this space is controlled environmentally: it is never too hot or too cold, the air always circulates, and it is policed, patrolled by uniformed security personnel, overseen by surveillance cameras, encircled by a small number of walled entranceways and regulated by a central management authority (see figure 4.7). Nothing that is unplanned can happen here. In theory at least, no crime or vagrancy or unpleasantness will occur—despite the vast numbers of people who move through the centre each day—and where it does (as in bagsnatching or a brawl), there are routine procedures to deal with it.

The postmodern shopping centre is thus increasingly a place for fantasy and amusement, for leisure as well as retailing activity, as the distinction between the two is deliberately blurred. It also remains a place primarily reachable by road, where the physical environment is hermetically sealed and, crucially, where there is a form of social regulation which includes security guards, surveillance cameras, and expected norms of behaviour, which declares this to be a privately owned public space. It is not one which the users or occupants or leaseholders own. The retailers, especially the smaller businesses, are subject to high rents, centre levies and promotional campaigns, and a range of regulations which cover colour schemes, display formats, lighting and placement in the centre. The larger scale occupants, the big department stores or anchor shops such as Grace Brothers, David Jones and K-Mart, are far more powerful in this scenario, and their location partly determines pedestrian flow and the flow of shoppers to other, smaller retailers.

These retailing spaces have precursors in nineteenth-century galleria, shopping strips, department stores and arcades (see Kingston 1994; Reekie 1993). What makes then Westfield Parramatta a postmodern rather than a modern space? There are four main distinguishing elements:

- the sheer scale of the centre and the range of activities it offers so that retailing space has been extended to include tourist and recreational space
- the design of the centre with its huge monolithic, inward looking structures, anchor stores linked by walkways full of smaller shops and glass-topped gallerias surrounded by a moat (or elevated banks) of car parking spaces
- the comprehensive regulatory and security control exercised by the centre owner–manager
- the importance of lifestyle shopping and the connection of leisure to retailing. These are new features of shopping centres related to the constitution of identity through desire.

This final point links the shopping centre as a place for purchasing commodities to the constitution of multiple and gendered identities.

Shopping for Identities

Those who try to pinpoint the postmodern characteristics of the regional shopping centre have isolated a number of economic and design elements. In addition, geographers have drawn on work in sociology and cultural studies which sees the shopping centre as a place where particular identities are offered and variously constituted (Gardner and Sheppard 1989; Lunt and Livingstone 1992; Miller et al. 1998). As Rob Shields writes:

> In shopping malls...new modes of subjectivity, inter-personal relationships and models of social totality are being 'tried on', 'taken off' and 'displayed' in much the same way as one might shop for clothes. These are not the modernist spaces of goal-directed individuals and utopian projects. Rather, contemporary consumption sites are spaces of carnivalesque inversions...The multiple masks of the postmodern person...'wears many hats' (Shields 1992, p. i).

How this process occurs in Westfield Parramatta can be seen in some of the promotional material which the centre circulates. The image which confronts everyone who enters this—and all other Westfields across the nation—in the Winter of 1999 is a smiling, brown-eyed, blonde thirty-something woman in a roll neck, pink jumper and grey jacket who, with manicured hand inquisitively resting against her neck, beckons with the message, 'A new season of change in store now'. She appears alongside the newest innovation from the Westfield Marketing Division—the Loyalty Program based on a co-branded credit card with Visa and ANZ; if you acquire this credit card and use it in the complex you gain reward points and shopping vouchers which can only be recouped with participating Westfield retailers (Westfield Annual Report 1998) (see figure 4.8).

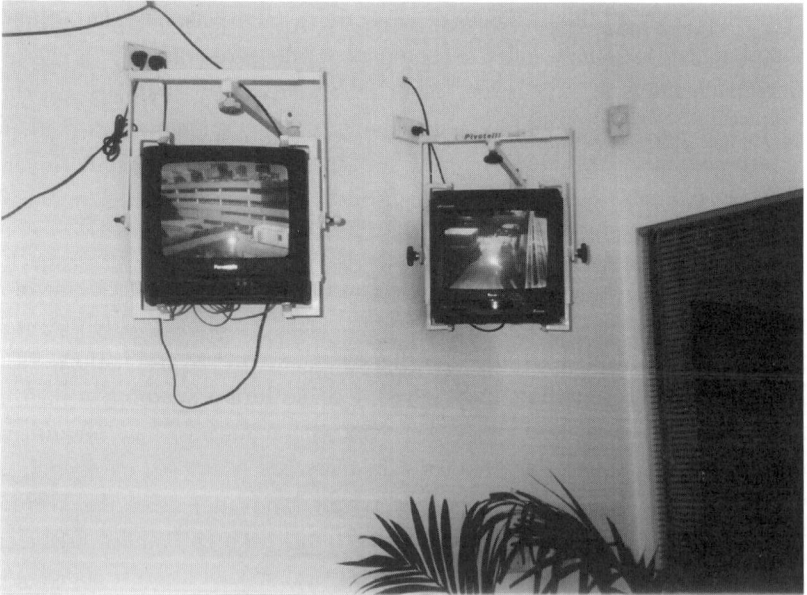

Figure 4.7 Elsewhereness, fortification and safety at Westfield Parramatta
Source: Louise Johnson

Figure 4.8 Buying change at Westfield, 1999
Source: Louise Johnson

The emphasis on fashion linked to a Western and Caucasian stereotype of female beauty and purchase continues with other promotional material in the store which similarly proclaims: 'Winter chic at Westfield'. In this region of ethnically diverse shoppers, the blonde woman of the posters reappears, this time in a red, fur-edged coat, carrying Westfield shopping bags, along with sixteen images (twelve women to three men, all but one obviously Caucasian) in a range of fashion wear, from neat casual jeans to silver cocktail dresses and Pierre Cardin suits. The back of the brochure notes that more than 160 (or 42%) of the nearly 380 shops in this centre supply fashion items. There are also twenty-six stores providing hair or health and beauty services, eighteen offering leisure goods and sixty-six (17%) serving or selling food.

What is being bought in such an environment? Writer John Berger argued in *Ways of Seeing*, that what is being sold in the shopping centre and associated advertising images is something that appeals very deeply to our sense of self and identity. He says that advertising creates a sense of inadequacy or lack, which is often sexualised, and which can only be eased through the purchase of whatever is being promoted (Berger 1972). In the shopping centre the connection between commodities and sexualised identities is made explicit.

Advertising and the presentation of, for example, clothing in a shopping centre window, relies on anxiety, insecurity and a fantasy-world where dreams can become real through the act of imagining and buying. Through such acts we become real and gendered in particular ways. Consider the window displays for Manpower, Table 8, Sussan or Cue Design (figure 4.9). They depict ideal men and women—even though the mannequins have no heads, only bodies!—who gaze out from behind the glass presenting a visage of style, availability and

Figure 4.9 Shop windows inside Westfield Parramatta, 1999—mobilising gendered desire
Source: Louise Johnson

desirability. The offer is that we, the gendered viewer/consumer, can become like these ideal men and women through purchase. For the woman, she is offered the image of the desirable, someone men will find attractive and other women will envy. For the man, the offer is one of power, style, success in business, in sexual conquest, in politics, and in the public realm.

What sort of desire is being generated in these images and shop displays? It would appear that browsing, and looking in the promotional brochures, at the centre posters as well as the store windows and in the stores themselves, and shopping become primarily an exercise in exploring and constituting a broad range of gendered identities. As Fiske, Hodge and Turner note:

> Image and style are not just means of self-expression, but of self-construction, and this self is constructed partly from and by the look...The action of the shopper is pleasurable partly because we see in the windows of the stores reflections of an ideal self that is potentially achievable through the purchase of commodities. Freud talks of the narcissistic function of the look (its dwelling on one's pleasure in one's own image) as well as the voyeuristic (the pleasure of possessing another through the look)...The power to see is the power to understand, which leads to the power to possess and control, and Freud's theory of voyeurism is specifically associated with masculinity (1987, p. 100).

Laçan extended Freud's theories to emphasise the difference between the imaginary and the real as the place where desire is born. For Rosalind Coward, it is female rather than male desire which is primarily mobilised in shopping centres and at a range of other sites. She comments that 'Women's bodies, and the messages which clothes can add, are the repository of the social definitions of sexuality. Men are neutral. Women are always the defined sex and the gyrations around women's clothing are part of the constant pressure towards display of these definitions' (Coward 1984, p. 30).

Such identities for women, who are still the main users of regional shopping centres, may indeed be fantasies but they can also be variously realised through the gaze or the purchase of commodities. If desire is for the exclusive fashion wear, the silver dress, the black suit or the slim and young figure, then the number of women who can make themselves into this image is relatively small. However, the number who can aspire to such images through projection, looking at fashion parades, other shoppers, shop windows, and actually trying on the outfits and vicarious purchase is far higher.

Research suggests that many women visit regional shopping centres to look rather than to purchase. For example, a 1978 survey of Westfield Shopping

Centre at Blacktown in Sydney indicated that 61% of visitors, who were mostly women, window-shopped or browsed, 28% watched other people and 27% watched an entertainment program (Department of Environment 1978, p. 7). Three-quarters of the mothers with young children spent more time at the centre than was needed to complete their shopping. In their work on Perth's Carrillon centre, Fiske, Hodge and Turner also confirm the importance, especially for young women, of looking at other shoppers, at windows, and at peers rather than buying (Fiske, Hodge and Turner 1987, p. 98). So clearly other things are occurring in shopping centres apart from shopping, one of which, I suggest, is a mobilisation of the female gaze and body into the world of fashion, even if temporarily. But a centre like Westfield Parramatta does not only offer the opportunity for fashion-related gender fantasies, it also presents possibilities for the realisation of such desires.

In the fashion brochures there are fashion items that are relatively affordable, that are worn by women over thirty, or who are larger than slight, svelte and waif-like adolescent girls. These fashion items can be tried on and bought. The credit card loyalty program makes the purchase of these items on credit easier; and this means that women move from the consumption of images and window displays into the stores which generate them, and from there to the items themselves, whether it is a David Jones, Jag, or a Table 8 version, or a Big W or Target item. In Westfield, the image is available at half the price at the discounter just down the corridor from DJs on Level 3!

Desire is thereby commodified but also variously realised through the acts of looking and fantasising, by donning masks and sometimes by buying. If gender can be so mobilised so too can class positions be defined, affirmed or challenged, at least symbolically, through shopping. As Daniel Miller and co-authors (1998) observe, after an exhaustive study of two north London shopping centres, class identity is something that can assume a specific form in a shopping centre:

> We do not encounter some free plurality of agents 'choosing' their class. We find instead a highly normative structure of oppositions...Class here is a practice, that is a processual relationship between people's consciousness and place. Its objectification in shopping is not unrelated to class as an expression of occupation, or of aspiration, but shopping provides a particular structure of difference that is not going to be quite the same as any other expression of class (1998, p. 187).

A similar set of observations could be made about gender differences and identities. While the choice offered by 160 clothing shops is immense, clearly only some will appeal, be explored, played with, tried on and bought. Which ones will be a complex function of identity discourses which intersect at the moment of window and actual shopping.

In addition to locating the play of identities within the world of fashion and clothing, which is clearly an important part of any shopping centre, there is also the growing emphasis in these centres on leisure activities and food. In particular, buying food is now linked to leisure as well as to family well-being, in that the woman shopping for the family is exhorted to buy fresh food, carefully and economically; she is also encouraged to sit, to linger and to eat in the many food courts and restaurants that are located in the regional shopping centre. I would argue that sites for the consumption of bought food are now critical points in regional shopping centres, as cafes, food courts and restaurants are increasingly located near entrances, at sites which assist in the flow through of shoppers from one precinct in the centre to another; and at particular places differentiated carefully by market segment. Thus take-away food bars for adolescents are located next to sports goods; quiet restaurants are located near homewares for older women and men; and fresh food retailers located near cosy cafes for women on their own. The food areas then are increasingly differentiated by age, gender and precinct. The food areas adjacent to a theatre or entertainment complex are very different in style and the age group for which they cater compared with food courts which are more comprehensive and oriented to families, and the cafes tucked away in quiet areas, or at key hubs of activity or scrutiny.

What sorts of identities are being mobilised and played out in such spaces? This relates both to the food and the precinct; but the identities are clearly identifiable, whether it is the hip young person going from INTENCITY to the take-away food bars, children imploring parents to move from the discounter and toy store to the hamburger chain, or the single person browsing for clothing who rests and watches in the secluded cafe. These fictional persons, who are the retailers' market segments, are usually seen as units: the young man, a child, a male or female thirty-something parent, or a twenty-year-old woman. But the spaces they occupy, the goods they buy and the foods they consume could be assumed by anyone who dons one of the roles the centre presents them. These are the possibilities opened by a postmodern idea of identity.

While the range of identity choices may be limited, postmodern theorists suggest the postmodern era presents numerous identities which are available to everyone. Westfield offers both gendered and embodied identities. The body, the original source of identity, is the subject of more than half the shops in Westfield Parramatta: the body can be clothed, fed, manicured, trimmed, exercised, entertained and constructed anew each day. In contrast, shopping used to be about getting the necessities of life as quickly and easily as possible—food, clothing and homewares. The range of shops was far more narrow and the task of shopping a more utilitarian one (Kingston 1994).

Geographers have seen in the postmodern a variety of changes which relate to the socio-spatial restructuring of whole cities and regions, the commodification and fortification of localities, and the constitution of new identities. The regional shopping centre is one space where many of these elements of economic, design and discursive change come together. For the feminist postmodern geographer, the playing out of desire, the constitution of sexed bodies in spaces such as Westfield Parramatta, and the deconstructive reading of such spaces against a background of their economic operation, allows the exploration of postmodern shopping. Other aspects of identity, such as class or ethnicity, are also present for those who shop, and are integral to the negotiation of identities within regional centres. In their study of two North London shopping complexes, Miller and his co-researchers found anxieties over racial and ethnic difference were critical to people's experience and perception of the two centres. This kind of emphasis is rare in postmodern feminist geographies, but Miller and his co-researchers tend to avoid the issue of gender difference! In the act of shopping, and in postmodern feminist geographies, what is promised, but rarely attained, is a multifaceted identity which is differentiated not just by a range of sexualities and gendered possibilities but also by ethnicity and race. This has been the focus of postcolonial geographies and is the subject of the next chapter.

5

Postcolonial Feminist Geographies

One of the fundamental insights in the postmodern critique of Western knowledges is its recognition that these knowledges are partial and fixed in time and space, and that, as feminists have pointed out, they have a masculine bias. At the same time, feminism has been subjected to a critique from women in marginal positions in Western feminism, that is, from indigenous women in colonised countries, migrant women in First World countries and women from non-industrialised economies. These postcolonial feminists asserted that white Western feminism was self-interested; and they now demand a voice for silenced women and a reconceptualisation of feminism around position, partiality and difference.

This postcolonial critique is particularly relevant to geography since the discipline was part of colonising processes. Its complicity began with geography's early fascination with 'Other(ed)' places; it was extended by the geographer's role in European 'exploration' and 'discovery' and in the regulatory procedures of cartography, social surveys and the classification of peoples in colonised countries. Thus one aspect of postcolonial geography details the role of the discipline in European imperial projects. In this, it reconstructs imperialism not only as a military invasion of a country, but as a set of regulatory practices and a colonisation of the mind whereby dominant Western powers impose foreign values and beliefs on 'Others' and imperial minds themselves also become shaped by the experience. Another aspect of the postcolonial critique seeks to include in the geographical canon statements from the colonised which are unmediated, not interpreted and untranslated. This critique of pre-existing knowledges has forced a re-examination of key concepts and frameworks. Consequently, postcolonial geography's understanding of spaces are substantially different from earlier frameworks.

In postcolonial feminist geography, this new understanding is both reflective about existing geographical work and extends postmodern feminism's emphasis on social differentiation and flexible identities. Thus positionality, difference, power and partiality become central; so much so that the idea of women as a unified group from which a united feminist politics emerges all but disappears. The final part of this chapter considers an example of this geography: a dialogue about Kooramindanjie Place or Carnarvon Gorge in Queensland, between Jane Jacobs, an Anglo-Celtic cultural geographer with a special interest in the Aboriginal sacred, and Jackie Huggins, an Aboriginal historian. The dialogue began as a conversation and appears here as it was later written. First, however, I outline the emergence of the postcolonial position and postcolonial feminism in order to locate the approach Huggins and Jacobs take.

POSTCOLONIALISM

Postcolonialism is a political and cultural movement associated with struggles for political independence. While military invasion and colonisation have a long history, it was not until the sixteenth century that European expansion reached the southern hemisphere, and not until the latter decades of the eighteenth century that the great southern voyages of 'discovery' were launched. By the nineteenth and twentieth centuries, Western powers, such as the United Kingdom, Spain, Portugal, France, Germany, and the Netherlands, had expanded into Africa, Asia, the Americas and Australasia. European colonisation first concentrated on the tropics where it re-oriented local populations and economies to trade or plantation agriculture. As a product of the 'age of discovery', Australia was initially invaded to guard these trade and plantation interests in Asia and India. The native population, which was numerically small, nomadic, racially distinct and geographically scattered, was defined out of existence by the British through the idea of *terra nullius*, the claim that Australia was an empty continent (Denoon 1983). However, the United Kingdom's rise as a global industrial and naval power drew Australia into a very different economy; Australia provided agricultural raw materials in return for manufactured goods, and became a place to send surplus population and to invest capital (Darian-Smith et al. 1996; Frost 1994; McCarty 1978). Capitalism was central, as Dirlik notes: 'Without capitalism as the foundation for European power and the motive force of its globalization, Eurocentrism would have been just another ethnocentrism' (Dirlik 1996: 307). In this military and economic regime, indigenous populations that could not be incorporated as a labour force, were forcibly displaced from the land, and sent to religious missions and Protectorates—created and administered by government and the church—where they became spatially and socially segregated (Huggins and Blake 1992).

Since the mid-nineteenth century and the growth of internationalisation and globalisation, the dominant process has been de-colonisation: European powers have either withdrawn or been forced to retreat from rebellious and autonomous states. These waves of settlement and retreat have been neither uniform nor universally exploitative. In answer to the question, 'When did the postcolonial begin?', Ella Shohat points to 1901 as the date when the Australian colonies federated and gained some measure of political independence, to the 1930s when Iraq became autonomous, the 1960s when the French withdrew from Algeria, and the 1970s when Angolans attained independence. She concludes that the meaning of de-colonisation varies between countries where European settler populations negotiated political autonomy, and others where the indigenous populations struggled violently and won independence (Shohat 1996).

While the timing of colonisation is highly variable, so too are the processes and outcomes of de-colonisation. For example, in the 'settler societies' of Canada, Australia, New Zealand and South Africa, European populations came, saw, conquered and stayed on to dominate and enrich themselves but not the indigenous populations. In contrast, for example, in India and Africa, colonisation involved the calculated incorporation of the two societies, and a withdrawal by the colonial powers which often resulted in political and economic chaos. Thus Stasiulis and Yuval-Davis distinguish between settler societies and colonies of exploitation. Settler colonies, such as Australia, were characterised by a relatively large European population of both sexes who depended on their imperial sponsors even as they attained economic and political autonomy in the land they settled. In contrast, colonies of exploitation involved the appropriation of land, natural resources, and labour as well as indirect control via a 'thin white line' by sojourning male administrators, merchants, soldiers and missionaries (Stasuilis and Yuval-Davis 1995).

These two waves of colonisation and de-colonisation have continued to shape virtually all continents and populations in some way. Postcolonial thought recognises this process as central to all countries, populations, and cultural formations.

In postcolonial writing, imperialism not only involves military invasion and economic integration but is associated with, and dependent on social and cultural transformations. For imperialism to occur, Europeans had to assert themselves as ideologically, militarily, and economically superior to the rest of the world. Associated with a set of Eurocentric, Christian, and racialised views around whiteness, industrial 'development', and Christianity, Europeans were encouraged by a range of discourses as much as by economic imperatives to venture beyond their borders in search of adventure, wealth, conquest, and religious conversions. In this framework Europeans sought to understand, appropriate and then make over in their own country's image the countries and societies they colonised. In his seminal book *Orientalism*, Edward Said (1979) shows how Westerners constructed a unified fantasy of the Orient both to justify and transform the Asian countries they invaded. Said suggests this image was also vital to the construction of 'Europe' as a unified, dominant and superior continent and society.

Thus colonisation required a redefinition of European society. Imperialism also meant the creation of deep psychological relations of dependence and hegemony. For both the imperial powers and the colonised, this was an incomplete endeavour. Homi Bhabha suggests the colonised populations mimicked or parodied the behaviour the imperial power expected, thereby undermining its effectiveness and power (Bhabha 1990, 1994). Rey Chow argues that far from the European gaze being completely dominant, the coloniser felt looked at—gazed upon—by the native, making the coloniser 'conscious'. This self-consciousness, derived from engaging with an active, challenging indigenous population, which in turn, produced the coloniser as a subject. The imperial subject exercised the potent gaze, but it was a gaze built upon the active questioning of the 'native'. The constitution of identities within a colonised society was thus predicated upon the imperial relationship and the racial stereotypes which accompanied it, both of the 'natives' and their 'masters' (Chow 1996; also 1989).

These processes of colonisation and de-colonisation thus created a set of economic, military and subject relations in the imperial and colonised countries; they also displaced huge populations, producing global diasporas and fractured identities. These movements propel Pakistanis to London, Koreans to Los Angeles and Vietnamese to Paris where they, in turn, intermarry and interconnect with other migrant groups. The results are complex societies where the category of 'migrant' and 'native' becomes blurred, and where identity is constituted along a myriad of national, ethnic and racial axes all built upon the vortices of imperialism (Bhabha 1994; Hall 1990, 1991, 1993).

Colonisation and de-colonisation are therefore processes which embrace military expansion, economic transformation, population displacements and identity reformations. Defining the process in these broad terms and pitting it against Western academic discourses are fundamental aspects of the postcolonial position.

Postcolonial Discourse

Postcolonial discourses have arisen primarily from those societies which have been subjected to the subordinating power of European colonisation. These discourses include an analysis of and resistance to that process, the views of indigenous and settler peoples, the voices of migrants displaced from their homelands by imperialism, war, poverty and enforced labour, and of those who have reflected on this process from within Western societies.

It is no accident that such discourses emerged at a time when formal colonisation had almost ended, when globalisation was gathering pace and it was clear that new forms of colonisation were replacing earlier forms. Postcolonial critiques also emerged at a time when newly industrialising Asian countries were challenging the economic hegemony of the West. Thus postcolonialism has a material foundation in globalisation and regional economic realignments as well as in formal de-colonisation. Despite its origins, postcolonialism has primarily been a way of thinking—a discursive rather than a political movement, with its focus on researching and writing the colonial and migrant experience, and its emphasis on giving voices to those previously silenced. In Western countries postcolonialism is not associated with political agitation for rights and compensation but rather can be seen as part of the postmodern cultural and linguistic turn in academic circles. As a consequence of this, postcolonialism for some is a deradicalising gesture, an evacuation of a crucial political agenda (Appiah 1996; Dirlik 1996).

The origins of postcolonial thought in the academy can be traced to the idea of the empire writing—rather than striking—back in the 1970s, when writers from Commonwealth countries demanded that their work be included in the canon of 'English literature'. Thus postcolonialism in the academy is an outgrowth of what formerly was called Commonwealth literary studies, a canon which came into being after 'English' studies had been broadened to include firstly American, and then other national and regional literatures. This was how English Departments engaged with the postmodern concept of 'difference', by seeking to broaden but not destroy the idea of English literature (Slemon 1996, p 74).

Postcolonialism also emerged from writers, historians and activists in 'Third World' cultures, such as Fanon (1968), who described their experience of colonisation. One of the most influential has been the Sub-Altern Studies Group in India. Its members have worked at recovering the history of anti-imperialist, political and cultural struggles by marginalised groups as well as at writing the history of India from the indigenous side of the imperial experience (Guha 1982–87; Prakesh 1990; Spivak 1987, 1990).

Others who have contributed to postcolonialism include migrant populations who are now writing of their experiences as colonised, displaced and marginalised peoples, and whom Stuart Hall describes as hybrid and diasporic communities. Hall sees this migrant experience as emblematic of the era, in that personal identity is unclear and temporary, formed at what he calls 'the unstable point where the "unspeakable" stories of subjectivity meet the narratives of history' (Hall 1987, p. 44). It is from this position that notions of race and ethnicity emerge to create new identities and politics, and to raise fundamental questions about the decline of the West and the marginalisation of all those who live within it (Hall 1987). The result is a burgeoning literature from migrant populations in North America, the United Kingdom, Europe and Australia who, as the displaced, write of their migratory experiences, their ambivalent identities and the racism in their new countries (Hall 1990, 1992; 1993).

In addition to those voices from colonised societies and from migrant populations forced to move within empires, postcolonialism as a critical discourse has involved subjecting Western academic knowledge to rigorous critiques. Thus there have been exposés of the complicity of anthropology and geography in the acts of 'discovery', mapping and colonisation; of political science and sociology in the ongoing regulation of colonised populations; and of history because of its erasure of the horrors of ongoing colonisation in the creation of national histories and identities. These critiques have included theoretical traditions in the humanities and social sciences and their complicity in the imperial project. The postcolonial discourse thus claims to go beyond the tenets of postmodernism, despite a view that postmodernism is one of the theoretical foundations of postcolonialism, since it challenges Western and Enlightenment hegemony, recognises difference and celebrates multiple identities. This claim—that postcolonialism goes beyond postmodernism—is made on the supposition that postmodernism, with its focus on texts and their relativity, is unable to engage with the power dynamics which maintain existing colonial practices (Triffin 1990; Ahmad 1995).

Postcolonial discourse thus points to the centrality of colonisation in the formation of Western cultures. Postcolonialism embraces voices from the colonies which describe the process of imperialism and the struggles for decolonisation; it includes work by those displaced as migrants, as well as the writing of those who are living in societies and working in academic fields created by these power dynamics and population movements.

Australian Postcolonialism

Australian postcolonialism builds on this overseas work but also arises from the particular conditions of imperialism and ongoing migration experienced by this 'settler society'. The form and timing of Australia's postcolonial critique emerges from both a practice of resistance by indigenous Australians and a politics of difference demanded by non-English speaking migrants. Recovering and publishing the violent history of colonisation has forced a major rewriting of Australian history. This has meant abandoning what the anthropologist Stanner, in the 1960s, called 'the great cult of forgetting', whereby Australian history was a story of heroic European settlement. It has been replaced by a view which sees the subjugation, murder, and resistance of Aboriginal population as critical, if not central, to Australia's colonial experience and foundation. In revising Australian history, Charles Rowley's and Henry Reynolds's research and writing on Aboriginal–White contact, resistance and interdependence has been vital (Reynolds 1972, 1990a and b; Rowley 1970, 1971). Their writing exists alongside and draws on that from Koorie, Murri, Nyunga and Nungga people who elaborate their encounters with kin, place, resistance and racism (Bandler 1989; Gilbert 1977; Langford 1988; Langford Ginibi 1994; Morgan 1987; Narogin 1990). This autobiographical work affirms colonisation as both a historical process and a continuing one. It also suggests that Australia is not yet a postcolonial society, given that indigenous autonomy is far from attained and struggles for even small rights over land, resources and rights continue.

Jackie Huggins, for example, details the way colonisation continued well beyond the early frontier encounters. She describes how the system of reserves, schools and missions established in the nineteenth century, tore families apart—as children were forcibly removed from their parents—imposed alien ideas of hygiene and Christianity, and created a labour force for pastoralists, both in their homes and on the land. She argues that such practices continued well beyond the era when the regulatory activities of welfare agencies and police sought to disband these institutions (Huggins 1987a, with Thom Blake 1992).

Postcolonial writers in Australia also embrace calls from non-English speaking migrants for a voice within a dominant Anglo-Celtic culture and a critique of the monoculturalism of that culture. Thus Sneja Gunew presents 'multicultural writing' as a challenge to British culture in Australia. Through the creation of bibliographies, anthologies and specialist libraries, the work of these writers becomes visible (Gunew and O'Longley 1988; Gunew and Mahyuddin 1992). Gunew's critical writings offer further challenges; for example, her 1990 essay on multicultural readings of Australia juxtaposes a retelling of the standard and authorised history of Australian writing against excerpts from non-Anglo Celtic women writers who have been denied a place in the canon (Gunew 1983, 1990). Elsewhere, she questions the ideas of centre, and margins, and irreducible differences between migrant and Australian-born writers (Gunew 1993). Here Gunew joins Ien Ang in suggesting that postcolonialism may involve the tolerance of difference by dominant groups, that is, the safe incorporation of diversity under the guise of multicultural difference when, in fact, ethnic and racial differences may be irreconcilable (Ang 1995). Thus Ang argues that: 'To focus on resolving differences between women as the ultimate aim of "dealing with difference" would mean their containment in an inclusive, encompassing structure which itself remains uninterrogated' (Ang 1995, p. 60).

Ang and Gunew join Jan Pettman in deconstructing the categories used in discussions of demographic and cultural difference in Australia. Thus Pettman surveys the contested and highly political use of the terms 'Aboriginal' and 'migrant' to challenge notions of fixed identity, and offers a future project that is feminist, antiracist and anticolonial (Johnson 1994c, p. 144):

> Central to this project is how to treat difference in theory and practice, and how to talk and work across the category boundaries. It means owning one's own speaking position/s, social interests and politics, and locating multiple and contingent identities and politics within structures and social relations. It means finding a language to interrogate difference, secure home bases and build alliances in a critical engagement (Pettman 1992, p. 58).

Postcolonial theory and practice in Australia therefore involves rewriting history as an ongoing imperialist project; admitting the voices of those silenced, incarcerated and marginalised by that history; and a politics of resistance, difference, visibility, disruption, and positionality by those of Aboriginal, Torres Strait Islander and non-English speaking background. In this exercise, Australia's demographics and experience joins with overseas postcolonial literatures to create a particular postcolonial discourse. In Australia and elsewhere, this discourse is given a further edge where it engages with feminism.

Postcolonial Feminism

In 1976, Aboriginal activist, and magistrate-to-be, Pat O'Shane wrote:

> The problem of racism is one that all women in the women's movement must start to come to terms with...it is necessary for (the) women involved to examine carefully whether or not their aims as white women are neces-

sarily those of black women...it appears to me that, whereas, for the majority of women involved in the women's movement, sexism is what the fight is all about, for Aboriginal women—when they look at all the medical, housing, education, employment and legal statistics—it becomes very clear that our major fight is against racism...And when the white women's movement takes head on the struggle against racism, which is the greatest barrier to our progress, then we've got a chance of achieving sisterhood (O'Shane 1976, 32–4).

In the eyes of many Aboriginal women, white women's dominance of the women's movement meant that feminism addressed only European problems. White feminism was limited, and posed additional problems for a group already crushed by over 200 years of colonisation. Aboriginal women accused white feminists of a profound and disabling racism which had had major consequences for Aboriginal women in Australia. For example, Bobbi Sykes observed how the white women's campaign for abortion on demand raised many dilemmas and dangers for Aboriginal women who had long been subjected to coerced abortion and enforced sterilisation. Indeed, Sykes argued for stricter controls not greater liberalisation (Sykes 1989). In a similar vein, Jackie Huggins has pointed to the self-interested and theoretically absurd feminism which, in the name of 'sisterhood', equates upper-class white women with Black women in the name of 'sisterhood'. She also affirms the colonial and racist nature of Australian society, and argues that Aboriginal women need to ally with black men in the struggle while simultaneously recognising their sexism (Huggins 1987a; see also Langton 1988, 1990).

These observations and claims echo those made elsewhere. Together they represent a profound shift in the nature of feminism. Thus if 'difference' for feminists in the 1960s and 70s meant those between women and men, in the 1970s and 80s it broadened to include class, race and sexuality. However, as Elaine Jeffreys notes, these admissions were 'add-ons' since white feminism remained the central point around which 'difference' was assessed and admitted (see chapters 1 and 2). A woman experienced oppression primarily because she was a woman, and only secondly because she was black, working-class or lesbian. In this conceptualisation 'sisterhood' remained inclusive, diverse, tolerant but also dominated by white, Western and middle-class women (Jeffreys 1991). Thus the challenge in the new century is to fully acknowledge 'difference' between women. As Anna Yeatman notes, the critique has to go much further and actually destabilise the categories on which Western feminism (and liberal humanism) is based: 'For a consciousness of multiple and interlocking oppressions to be possible, the idea of universal emancipation has to lose legitimacy. This occurred with the effective challenges in the postcolonial era to the assimi-lationist character of the universal or humanist subject' (Yeatman 1995, p. 51).

Such challenges came first from feminism and then from postcolonial challenges in feminism. Thus, in the early 1980s, the history of oppression, activism and marginalisation of Black and minority women in the USA, the United Kingdom, Australia, and New Zealand erupted into a series of critiques of white, Western feminism. In the USA, a collection entitled *This Bridge Called My Back: Writing by Radical Women of Color* made acerbic comments on 'racism in the women's movement', and sought to 'reflect an uncompromised definition of feminism by women of color' by discussing the relations between culture, class, the Third World and homophobia (Moraga and Anzaldua 1981). A range of writers noted white feminism's misrepresentation, homogenisation and ignorance about other women and the long-term connection of imperialism and racism to feminism. Their calls ranged from urging white women to listen to their voices (Carby 1982; Mani 1990) to stinging indictments of their exclusion and marginalisation from white societies in the USA and the United Kingdom (Davis 1981; Minh-ha 1989).

In response, many white feminists engaged with the issue of racial difference; but this also proved fraught and spurred critiques of the ways in which Black and Third World feminism had been homogenised and included in white feminism (Amos and Parmar 1984; Chow 1996; Mohanty 1988; Ong 1988). For example, Chandra Mohanty examined a series of texts published by Zed Press in the 1980s which purported to admit women into studies of 'Third World development'. Her analysis revealed that many of these texts implied women were a coherent group with identical interests and desires (regardless of their class, ethnicity or race), that they offered dubious 'proof' of such universality, and portrayed the Third World woman as poor, ignorant, traditional and domestic and Western women as wealthy, educated and modern (Mohanty 1988).

Apart from feminism's exclusion of some women, Third World and indigenous women noted many points of conceptual marginalisation. Those writing from 'minority' positions as indigenous women or women who were ethnically differentiated from a dominant population, began systematically deconstructing key concepts, thereby revealing the boundedness of feminism. They examined terms such as 'patriarchy', 'family', 'production', 'reproduction' and 'sexuality' and found that their definitions were largely white and Western. One deconstruction led two white British socialist feminists, Michele Barrett and Mary McIntosh, to reconsider the views they expressed in their book, *The Anti-Social Family*. There, Barrett and McIntosh had argued that the family has a debilitating effect on (all) women (1982). For many Black and Asian women in the United Kingdom, however, 'the family' is one of the few positive places they occupy, a setting that extends widely, a key foundation of their communities and thus structurally quite different from the family forms McIntosh and Barrett were describing. Barrett and McIntosh came to see that the often dysfunctional nature of Anglo-Celtic nuclear families was not typical of Black and Asian families; the latter were the sites of strength and resistance, the institution on which many battles were waged with immigration and social welfare authorities. Barrett and McIntosh's book devalued these struggles and the strength of the extended family in Black and Asian communities. Some described Barrett and McIntosh's efforts at reconceptualising 'the family' as an institution in which ethnicity makes a difference (Barrett and McIntosh 1982, 1985), as complacent, patronising, defensive, overly generalised, and tied to the demands of socialist-feminist theory rather than those of Black women, and subsumed by a mild politics of ethnic difference rather than by any serious understanding of the pain, divisiveness and theoretical challenges which racism poses (Kazi 1986; Mirza 1986; Ramazanolglu 1986). These exchanges signalled the fraught politics surrounding an engagement with difference.

The late 1980s thus brought a questioning and a reconceptualisation of feminism. Add-on models of oppression were no longer deemed adequate to deal with the demands of difference; feminists have been forced to rethink some of their key assumptions, categories and priorities. The new feminism is still emerging; it accepts many of the postmodern insights on identities as positioned in both a national and transnational frame and points to the interrelationship of oppressions of race, class, ethnicity, gender and sexuality (Clough 1995, p. 118). Thus, as Teresa de Lauretis notes, Black women in the United Kingdom experience racism not as 'Blacks' but as Black women; thus the female subject is multiply organised along several social axes with mutually contradictory discourses and practices (de Lauretis 1988).

There have been other, new ways of viewing and writing women's difference. In a book which challenges a range of discursive power relations, Gloria Anzaldua describes living in and through the borderlands between the USA and Mexico. Through a mixture of Spanish and English prose and poetry she evokes the hybridity of living in and between multiple worlds of race, ethnicity, language and sexuality, worlds which are not only enriched by this diversity but structured by the relations of power which attach to each (Anzaldua 1987).

In Australia, these multiple oppressions are also mediated by an overwhelming colonial experience. As Jackie Huggins notes: 'Australia was colonized on a racially imperialistic base and not on a sexually imperialistic base. No degree of patriarchal bonding between white male colonizers and Aboriginal men overshadowed white racial imperialism. White racial imperialism gave all white women the right to oppress Blacks—women and men' (Huggins 1987a).

Thus postcolonial feminism must acknowledge not only multiple oppressions and identities but also the realities of dominant power dynamics which are specific to times and places. With its focus on the relationship of people to their place, geography becomes critical to the realisation of the postcolonial critique.

POSTCOLONIAL GEOGRAPHY

Postcolonial geography builds on postcolonial discursive and political movements across the world. However, as a discipline, geography has assumed a special status which derives from the complicity of its members in colonising processes, and from the importance that notions of postionality and location have assumed in the creation of postcolonial alternatives. As a result the emerging postcolonial geography appears to have four main elements:
- detailing the complicity of the discipline in continuing processes of colonisation
- questioning Western and geographical representations of its landscapes
- admitting the transforming voices of the colonised into the discipline while also recognising their diversity, ambiguity, hybridity and multiple identities
- bringing to the fore the politics of positionality in writing postcolonial geographies.

Feminist geographers continue to assert the importance of gender, but this is tempered by an emphasis on the centrality of multiple, contingent and strategic identities and power relations. This modifies the former focus of feminist geography which, from its inception, was on women as women. Questions of patriarchal power and gender relations are therefore to be asked and answered within a broader frame of multiple axes of difference. Jane Jacobs and Jackie Huggins clarify the meaning of this approach as they discourse on one place in Central Queensland. But first, their postcolonial feminist geography needs to be located within the larger enterprise of the discipline.

Geography and Colonisation

Geographers have begun to reveal how the curiosity which instigated the discipline was also implicated in the colonising process, and how mapping was crucial to the European voyages of 'discovery' and to their colonisation of these countries (Harley 1989; Wood 1992). As Duncan and Sharp (1993) note, the representation of 'other' people and places necessarily involves a power relation mediated through various cultural institutions and practices, through class and gender formations and through taken-for-granted assumptions about history and progress. Interest in other places can thus be seen as evidence of unequal power relations, ones which involve acts of imagination as much as invasion (Duncan and Sharp 1993; Duncan 1993; Gregory 1994). The exercise of this power by geographers primarily took the form of mapping, describing and naming places.

The power of mapping has been linked to that of naming and the imperial gaze. Thus the naming of Australian landscape features in English and the creation of maps with large areas of blank space suggesting no occupancy, were imperial rather than scientific acts; these acts discursively obliterated an Aboriginal presence and indigenous landscape relations and names (Crush 1994; Driver 1992; Johnson 1994c). In a spatial history of an outer suburban estate, I have written how the earliest maps of Melbourne included features of significance to European occupiers but left blank those areas where Aboriginal people were known to be (Johnson 1994c and d). Similarly, the historian Tony Birch has traced a related process of mapping the Grampians (or Gariwerd) in north-western Victoria. He outlines the controversy over renaming parts of this landscape with Aboriginal names, as an act of white indulgence aimed at a tourist image and conducted without the involvement of local Aboriginal people. He therefore sees this conflict as a contemporary example of a continuing process of colonisation (Birch 1996).

The geographer's role in colonisation went beyond the making of maps and the naming of places, to include the construction of 'othered' places through forms of description which privileged the European gaze and facilitated military or settler occupancy. Thus Felix Driver notes how notions such as 'darkest Africa' were created, circulated and read to imbue the European imagination with a particular set of racialised, stereotyped, homogenised and negative connotations (Driver 1992). So too, May Louise Pratt details how travel writing, diaries and scientific accounts by Europeans of their 'discoveries' in Asia, Africa or America involved imposing an urban, middle-class and white sensibility upon landscapes and people. It was, as Pratt (1992) suggests, following Said, the creation of a European fantasy which had devastating consequences for the territories and peoples surveyed (Pratt 1992). From these historical reconstructions of past explorations, Hudson observes how the geography of the late nineteenth century was promoted to serve the interests of imperialism which included territorial acquisition, economic exploitation, militarism, and the practice of class and race domination (Hudson 1977).

As well as examining the role of the discipline in historical acts of 'discovery', invasion and settlement, geographers have also considered continuing processes of colonisation. Thus David Harvey, Robert Newman and Neil Smith tell of

Owen Lattimore, a historical geographer and east Asian specialist and Director of the Walter Hines Page School of International Relations at John Hopkins University, who, in the 1950s, advocated a strong anti-racist and anti-imperialist position. Following this work, he was named by Joseph McCarthy as 'the top Soviet spy' in the USA, denounced by fellow geographers and rendered *persona non grata* in the academic world (Harvey 1983; Newman 1992; Smith 1994). Smith explores the saga of geographers who actively assisted US imperialism by describing how Harvard University produced informants and trained people for the CIA and FBI (Smith 1994; see also Diamond 1992 and Livingstone 1992). So too, Jonathan Crush describes the dilemmas faced by geographers in South Africa as they attempt to decolonise their own discipline, an exercise with an immediate political urgency and theoretical relevance (Crush 1994). The fact that there can be an empire 'at home' as well as overseas, and that such imperial relations do not belong solely in the past, are key elements of postcolonial geography.

Following this work on the discipline's complicity in continuing colonisation, postcolonial geographers such as Peake (1993) and Smith (1994), have lambasted existing studies which ignore racism and ethnic difference, and have also noted the dominance of white members of staff in departments of geography in Western universities (Coher 1990). Others, such as Rickie Saunders (1990), view efforts to admit the postcolonial position into the discipline and to transform it, as tokenistic and ineffectual.

Nevertheless, the process has begun of rewriting the discipline from a postcolonial perspective so as to include the colonial experience as central and to acknowledge the ongoing power dynamics resulting from racism, migration and global capitalism (Jacobs 1996; Keith and Pile 1993; Massey 1994; Radcliffe 1994; Rose 1995; Sibley 1995). In this exercise, the work of Doreen Massey in the United Kingdom and of Jane Jacobs in Australia has been central.

Massey has examined north London for the way in which it registers waves of investment, production and consumption relations as well as gender and race-specific migrations which accompanied the expansion, contraction and then the re-integration of the British Empire. She argues that places are always constructed out of social relations which are not only internal to that locale but which link them to elsewhere; she concludes that 'local uniqueness' is always a product in part of global forces (Massey 1994). In her work on both the United Kingdom and Australia, Jane Jacobs also takes imperial relations as central. Thus in her studies of inner city London, she outlines the ways in which the British Empire created both the financial magnet of The City and the ethnicities, class relations and heritage values which jostle for control in the redevelopment of inner city precincts. In contrast, Jacobs's studies of Perth and Brisbane in Australia admit the centrality not so much of contemporary waves of migration but of the ongoing importance of the white settler invasion of Aboriginal lands, and the subsequent commodification of those lands in the capitalist economy. Her studies reveal the consequent conflicts over urban spaces as products of imperial relations which are both historical and contemporary (Jacobs 1996).

Jacobs, Massey and other postcolonial geographers often mention the role of gender and that of imperial, class, ethnic and race relations in creating colonial and postcolonial spaces. However, they rarely make gender central. Rather, they position women and men in a web of power relations and social dimensions which tend to privilege either multiple identities or the politics of imperialism. Whether feminist geography can be located in such a discourse is a difficult question.

Postcolonial Feminist Geography

Despite the rhetoric of inclusiveness and the context of a postmodern feminism of difference, a number of feminist geographers have noted how little actual work has been done on race and ethnicity in the subdiscipline. Thus in 1990, both Liz Bondi and Rickie Saunders observed how the interrelations between class and gender had received considerable attention in contemporary geography whereas issues of race and ethnicity had been ignored (Bondi 1990; Saunders 1990). Linda McDowell and Susan Christopherson have observed how pre-existing feminist geography has been challenged by Black women in the United Kingdom and the USA, by critical voices emanating from the 'Third World' as well as by deconstruction. They argue that it is time to respond creatively (Christopherson 1989; McDowell 1991b). The call to recognise racial and ethnic differences is familiar in feminist geography circles but, as Brenda Yeoh and Chang Tou Chuand point out, if feminist geography is to 'ensure that spatial and social otherness is acknowledged, there is a need to refocus substantive research from "glamorous" first World sites to the "gritty" realities of the Third World' (1995, pp. 126–7). They suggest that such a shift would mean decentring academic interests and political alliances, identifying new sub-disciplines and practitioners, and 'opening up new ways of representing and communicating with Third World men and women which neither entrap them within a Eurocentric mould [n]or grant them a privileged position as 'authentic experiences' (1995, p. 127). It is this task that postcolonial feminist geographers have taken up.

In order to grapple with the issues the postcolonial critique raises, feminist geographers such as Sarah Radcliffe have developed a range of strategies. These include rewriting male-centred 'development geography', serious consideration of the critiques coming from black and Third World women, and the creation of authoritative spaces by those women in positions of discursive power so that marginalised women and men can have their voices heard (Radcliffe 1994). However, as Mohanty (1988) has noted, it is neither easy to include women in development geography, nor to admit the postcolonial critique when it becomes an add-on rather than a transformative exercise. The postcolonial critique involves altering geography in general, including feminist geography. Such a transformation means listening to 'other' voices, recognising the position and power of Western women, and allowing the destabilisation of geography and its central categories and assumptions.

According to Linda Peake, while postmodern geography explicitly argues for a recognition of difference, it has not substantially challenged the heterosexist and white cultural constructions which pervade geographical discourse. Peake then draws on the work of women of colour, and radical and lesbian feminists to rethink her research project, her relation to it and its outcomes. Thus she examines the everyday activities of two groups of low income women in Grand Rapids, Michigan, showing both their similarities in income and gender status, and their particularities which are primarily racial (Peake 1993). Peake notes the more complex ways in which women now have to be imagined: as points of intersection in a range of discourses and power relations about class, race and ethnicity as well as gender. The precise relationships between these elements cannot be predicted; although Peake maintains a concept of patriarchy, it is contingent on the particular circumstances in which the home–work dynamic is negotiated.

Mary Louise Pratt and Mona Domosh also keep women to the fore when they consider how women explorers and travel writers entered the imperial project compared with their male counterparts (Pratt 1992; Domosh 1991). In one of the few major collections of postcolonial feminist geography, Alison Blunt and Gillian Rose draw on a range of countries and times, including nineteenth-century Africa and Australia, and contemporary South Africa and Ireland, to put women at the centre of their studies while also recognising the multiple nature of women's identities, as well as their implication in the colonial project as it was variously negotiated across the globe (Blunt and Rose 1994). This has led to a range of different feminist geographies, which do not necessarily privilege gender but which keep women clearly in focus. This work recognises that women inhabit more than one sociological dimension: that is, a contingent range of discursive positions which always include the colonial and hence race, ethnicity and class (see also Fincher and Jacobs 1998).

How these geographies are to be researched and written raises critical questions about 'fieldwork', observing 'others' and presenting the results. In seeking to include these considerations here, I approached the Aboriginal activist Jackie Huggins. I had worked before with Jackie Huggins in creating material for teaching Women's Studies but before she agreed to take part in this project, she raised vital questions about the nature of the project and about my assumptions. These questions centred around the nature of identity: I had assumed only an Aboriginal woman could write this section, an assumption which Jackie pointed out contained elements of essentialism and escapism. As a historian, Jackie was uneasy about writing for a geography audience, an anxiety which led to her asking to be a co-writer. An accomplished academic, Jane Jacobs had worked with Aboriginal people for a long time, and was then, as now, deeply engaged with the problem of what a postcolonial geography might be. As she and Jackie talked about writing this section, they organised a trip to Carnarvon Gorge in Queensland, and considered the issue of how to express different authorial positions. The result is a section which writes a spatial history of this place from two, partial positions.

KOORAMINDANJIE PLACE

Jackie Huggins and Jane Jacobs

Jane

Unsettled authority and recent social and cultural theory have problematised the issue of speaking positions: it has questioned the political correctness or legitimacy of, say, a white Australian woman speaking or writing about the experiences of Aboriginal Australians (male or female). This marks a new self-reflexivity in the academy around the relationship between 'researcher' and 'researched' in relation to the politics of representation. This is a moment precipitated by challenges from both feminism and the politics of race (see as examples, hooks 1989, 1990; Minh-ha 1989). In short, there are challenges based around the politics of difference in which the 'other' has refused the gaze of the researcher, and reclaimed the right of self-representation.

There have been a number of responses from those displaced or challenged by this politics. Most dramatically, there has emerged the politically correct silence, a relinquishing of the 'right' to speak about the experiences of those previously positioned as 'other': that is, male researchers not speaking for women, white researchers not speaking for people of colour. The intention is that such silences create a space in which the sexed subaltern can speak (see Spivak 1987).

Another adjustment made to accommodate the politics of speaking, has been the proliferation of statements of self-positioning, the declaration of one's own position as, say, white-female-heterosexual, and the commitment to conduct research and writing within the limits inherent in that positioning.

Another response to this politics has involved a shift in the focus of research away from the experiences of the 'other' to the ways in which that 'other' is represented by one's own reference group. So, to take an example relevant here, for a non-Aboriginal researcher to shift from examining Aboriginal society to the ways that society has been represented by non-Aboriginal Australians. This then becomes more a study of whiteness (and by implication the privileged power of whiteness) and of the ways in which that whiteness is constituted, in part, through certain derogatory, restraining and racist renderings of people of colour. This argument parallels that levelled at masculinist science by feminism, in which the 'universal truths' of science are shown not only to be derived from masculinist knowledges but to depend upon sexist practices towards and renderings of women (see Rose 1993 for a discussion of this issue in relation to geographical thought).

The turn to representation or social constructionism has meant studies of race are more likely to consider, for example, the ways in which Aborigines have been depicted throughout the European history of Australia by sections of the settler community such as the State, missionaries or even anthropologists (see, for example, Attwood 1989). Such depictions are not impartial renderings of reality, but positioned, ethnocentric representations which often say as much about the settler–colonialist author as the Aboriginal subject. Studies which focus on the way in which Aborigines have been represented or socially constructed avoid the charge of presuming to speak on behalf of Aborigines, for they are actually speaking about settler society and its constitution of itself through the 'other'.

Social constructionist accounts appear then to be politically correct: they do not presume to speak for the other but speak of how the other is spoken about. However, their effect may be as troubling as more starkly invasive accounts of Aboriginal life. The critical distancing present in such accounts—often marked out by the use of quotation marks around derogatory descriptions—may be lost, and the effect may be to yet again give voice to racism and sexism. Take the case of a white, male university lecturer who, in his classes, carefully refused to speak about Aboriginal society directly, but talked instead of how Aborigines were seen by early white Australians. His politically correct intentions were construed by the class, and particularly Koori students in the class, as having a racist effect, despite his repeated gesturings of quotation marks around his reiterations of nineteenth-century racism. A similar controversy broke in Canada in response to the Royal Ontario Museum's exhibition 'Into the Heart of Africa' (see Jackson 1991a). What becomes silenced here is the critical inflection of the quotation mark.

These developments around the politics of speaking positions have important implications for the discipline of geography which has always held as vital, as one of its defining principles for the formation of knowledge, the practice of 'going into the field'. The issue of speaking positions has changed both the politics of entering 'the field' and what can be seen to constitute 'the field'; that is, what and who can legitimately and ethically be open to the geographical gaze (see England 1994; Katz 1994). What occurs in this moment of political reckoning is a rethinking of geographical reportage itself, a redefinition of the process of authorisation. This self-reflexive critique is apparent in other disciplines dependent on 'the field', such as anthropology (see the various contributions from the New Anthropology such as Clifford 1988; Clifford and Marcus 1986; Marcus 1992).

This section records an attempt to find a possible path through the politics of speaking positions. It does so by following the format of a tandem narrative: two writers not seeking to settle, through joint authorship, on a single narrative. Instead we are trying to demonstrate that there is not one story or, for that matter, name for Kooramindanjie/Carnarvon Gorge—there are many, two of which are presented here. The stories work with and against each other, they unsettle each other. The tandem narrative is not offered as the solution to the politics of speaking positions. Others, such as the anthropologist James Clifford, have offered alternative strategies of writing culture from a multiple perspective, and I think here of his essay entitled 'Identity in Mashpee' (Clifford 1988). To suggest there is one solution would be to diffuse what is a most important political process which should always remain negotiable, always be open to challenge. But this writing strategy is offered as just one way of rethinking the way in which the relationship between identity and place might continue to be written.

Jackie
Talking Back—Non-Aborigines must understand their cultural and ethical limitations in studying, researching and writing about Blacks; learn to step back in areas where they are not welcome; but often think and presume they are, where they are intruders rather than accomplices. Otherwise they do great damage to Aboriginal people and their struggles, adding to the burden rather than alleviating it.

Binney (1987) suggests that we cannot translate others' histories into our own, we can merely juxtapose them. To do otherwise is detrimental to the integrity of one or the other or to both historical traditions. Rather than trying to understand the past on its own terms, some academics have sought to explain

Aboriginal pasts in terms of a contemporary ethnographical present they confront but do not fully comprehend.

Knowing a socially constructed world is not the same as knowing it from within. The actualities of our everyday world are already socially and culturally organised. If we begin our understanding as we actually experience the world, it is at least possible to see that we are located, and that what we know of the 'other' is mutually conditional upon another location. There are different experiences of the world and different bases of experience. Whites must not ignore this by taking advantage of their privileged speaking positions to construct a sociological, external version which may pass for an Aboriginal 'reality'. One may not rewrite the other's world or impose upon it a conceptual framework which derives from one's own.

Jane

Geography and Colonialism—Geography and anthropology share a problematic relationship with the historical process of colonisation. Geographical knowledges of 'other' lands, like the anthropological knowledges of 'other' people, were crucial tools in the expansionist practices of colonialism. The making of maps, that quintessentially geographical practice, provided a practical guide for dispossessing 'others' of their place. The colonial history of nation states such as Australia testify to the symbolic and practical possession enacted through the map (Harley 1989; Livingstone 1992). The cartographic possession of Australia was premised by the conceptual emptying of the continent. Declaring the continent *terra nullius*, land unoccupied, was the conceptual violence that opened the way for the settling of Australia by the British (see Carter 1987). This initial act of violence has continued to be enacted through the history of Australia. The historical geographies of Kooramindanjie Place presented in this essay describe the ongoing violence of colonialism.

This process of colonial dispossession is explicitly gendered. Kay Schaffer (1988) argues that, in the early history of masculinist settler Australia, a woman's presence was registered through metaphors of landscape. Schaffer argues that Australian colonisation was of a land 'imagined, through metaphor, as the body of a woman', and that the designation of Australia as 'an empty space on the map of the world' presented it as a site open to, and given the attributes of, masculinist colonial desires (Schaffer 1988, pp. 22, 77–9). Schaffer makes explicit the link between masculine (man, Empire, Civilisation) and the subduing of the feminine (woman, Earth, Nature) in the settlement of Australia. The colonising of Australia is enacted through patriarchal constructions of masculinity and femininity in which the land and women were collapsed into a single category. The geographer, Janice Monk, has described a similar process in the North American colonial setting (Monk 1992).

If Australian land was feminised in the name of colonisation and exploitation, then the indigenous inhabitants of the land were in many renditions conveniently absorbed into that feminised nature. The declaration of Australia as *terra nullius* discursively emptied the nation. Early depictions of Aboriginal Australians often placed them as part of a feminised nature; sometimes passive, sometimes capricious or wild, but always to be invaded and possessed. But Aborigines were not eliminated, and in many parts of Australia they resisted the European invasion.

Aboriginal resistance and survival did not necessarily mean the masculinist, colonial gaze was avoided. For example, early anthropological accounts of 'traditional' Aboriginal society were translated through the lens of Western patriarchy. There was a lack of acknowledgment of, or a denigration of, 'women's business'; that is, the spiritual and ritual knowledges and practices managed by women. Male anthropologists either ignored the business of women or were denied access in accordance with the gender specific restrictions of Aboriginal society. The spiritual knowledge and ritual practices of men were often assumed to provide for the entire community, and women were viewed as 'profane', participating in 'small-time' rituals and magic unconnected with the more important issues of land and social harmony (Elkin 1939). It has only been through the work of feminist anthropologists, and the emergence of Aboriginal women politicians and historians, that this masculinist depiction of Aboriginal women has altered (see Bell 1983; Hamilton 1981; Huggins 1987b). These writers have challenged the view of Aboriginal women as 'feeders and breeders' servicing the loftier and more spiritual men. Aboriginal women are now known to have important land-based traditions, to be equally important as the men in maintaining the land, and to have much autonomy and power over the management of social relations.

Jackie

Colonising Aborigines, colonising women, the devastating impact of the European expropriation of land on traditional life, the introduction of European diseases, the officially sanctioned or unpunished murder of Aboriginal groups, the continuing efforts of governments, church and pastoral interests to destroy Aboriginal cultures and economies, the abuse and exploitation of Aboriginal women as chattels and slave workers—all this is ignored in Australian history.

Aboriginals adopted varied strategies for accommodation and survival. The eugenics movement characterised all Aboriginal women as 'morally feeble minded' and as requiring segregation in order to protect the white race. By the turn of the century, a number of 'protection' or 'welfare' administrative mechanisms were in place. Despite minor differences in policy, all eventually encroached on the independence Aborigines previously enjoyed.

They faced increasing control and confinement over the next fifty years, with intensifying intervention in communities and families, justified by the rhetoric of science and education. Many of these policies focused on women, who were variably seen as a means to increase an Aboriginal workforce; as the site for intervening in and controlling Aboriginal birthrates; or as cultural channels for the assimilationist policies which rejected Aboriginal values.

Assimilation would be achieved, it was believed, either if Aboriginal women were removed and 'educated' then returned to their communities, or if they were 'educated' while in their communities by being relentlessly inspected and evaluated by officials who demanded culturally inappropriate 'housekeeping' or child-care styles. The state, then, tried to use Aboriginal women against their own communities, either by separating them from their communities forever, or by trying to make them the instruments of cultural indoctrination.

Aboriginal women have been involved from the outset in Aboriginal struggles to regain land. Land Councils were a real attempt to establish culturally appropriate organisations, but they worked within European male paradigms and inevitably privileged men's knowledge and participation. Aboriginal women were, and continue to be, active in meetings about community affairs, but the Land Council's agenda was about the ceremonial and often sacred knowledge about land which, in Aboriginal society, is sex-segregated knowledge.

In many areas, women conduct ceremonial business which has parallel themes and goals to that of men, but it is not always appropriate or permissible for either sex to discuss their ceremonies in public or in mixed company. Although Aboriginal women's knowledge was valued by Aboriginal men and often proved essential to the outcome of land claims, women and their perspectives were at first excluded by the structure of the new land claim procedures.

Jackie

Kooramindanjie Place—My mother Rita says: 'We took a trip back to my born country recently. Tourism has taken its toll in the area, but the place still has its wild beauty. I felt the call of my people billowing through the trees and welcoming me home again. I saw the smiling faces of my elders, the embers of the campfire, heard the women singing. In my heart was such a deep happiness because I knew I was home again. "Rita Huggins was born somewhere out there in a cave," I said over and over again in my mind.'

Kooramindanjie Place is 600 kilometres north west of Brisbane in the sandstone tablelands of the Great Dividing Range. Its oasis in the desert plains was an attraction for Aboriginal people who would travel many hundreds of miles in search of water, food and shelter. It is also the country of Rita, my mother and my maternal grandmothers.

Our people lived in this area for over nineteen thousand years (archaeologically speaking) in the maze of gorges, ranges and tablelands. Some tribal groups living around the area were Kairi, Nuri, Karingbal, Longabulla, Jiman and Wadja. However, since the government removals policy, descendants from these tribal groups are now scattered over wide areas of Queensland and northern New South Wales.

Jane

Touring Kooramindanjie place in Carnarvon Gorge National Park, Central Queensland, a rock art trail guides visitors to an Aboriginal place in the past. The interpretative sign suggests that: 'Ten thousand years ago you may have followed a family group of Aboriginals across the dry plain to this oasis gorge' (Queensland National Parks and Wildlife Service [NPWS] 1991).

It is 1993 and I have just 'followed a family group of Aboriginals', The Huggins (Jackie, Aunty Rita and young John Henry), across the plain, to the oasis gorge and down this very trail.

The interpretative sign directs us to 'pause...reflect and wonder at the ways of those who wandered this land before us'. This family group and their trailing tourist wonder more about the ambiguous status of those who wander the land now. We discuss the history of violence and dispossession that resulted in the child Rita and her immediate family being removed from nearby Springsure to Cherbourg mission. We contemplate the history which has meant this visit is one of only a few ever made to the area by Jackie or Aunty Rita. I uncomfortably contemplate my first visit in the 1980s as an 'expert' on the management of rock art sites suffering the pressure of tourism and the meaning of this role in terms of the ongoing workings of colonialism. I resist the temptation to take notes, to take up the role of the ethnographer. I confine my note taking to the way the national parks depict Aboriginal associations with the area.

Jackie

Out of Place—I don't have to say to Jane that I didn't want her to be the big white expert in this exercise. In fact she knows well my feelings on that topic, but I can see she's itching to get to that spiral notepad. In fact my mother asks me 'Why isn't she taking notes?' She told everyone about Jane when we returned home, referring to her as another writer, like myself, but she must have been bewildered at her uncanny ability to take a book of mental notes.

I felt 'out of place' here, and Jane and I actually thought we were lost in the area at one stage. I felt I should know this place but could not get us out of the mess we'd found ourselves in. We did not know where we were as we drove along the dirt track in the darkness of night. I could sense that Jane was too frightened to offer directions—directions I had expected, her being a geographer and all. But despite getting lost, returning to my mother's 'born country' complemented my own sense and proudness of identity and sense of belonging. I began to gain an insight and understanding into her obvious attachment and relationship to her country and how our people had cared for this place way before park rangers ever clapped eyes on it. The way my mother moved around, kissed the earth and said her loud (but silent) prayers will have a lasting effect. She was paying homage and respect to her land and ancestors who had passed on long ago but whose presence we could feel there with a mutually knowing intensity.

This was our place, our sense of becoming. Like most Aboriginal people it is my spiritual and religious belief that we evolved from this land; hence the term 'the land my mother'. This land is our birthplace, our 'cradle of humanity', our connection with the creatures, trees, mountains, rivers and all living things. This is the place of my dreaming.

We don't believe we migrated south in the ice age as archaeologists and anthropologists reverently declare. There are no stories of migration in our Dreamtime stories, and science has never proven that we come from anywhere else except Australia. Our creation stories are intrinsically linked to the earth. This is why place and land are so important, to us it does not matter where and when we were born.

Jane

A Place Possessed—This is a place possessed with tourists, shaped by tourism. A place which, in my past role as an adviser on the management of rock art sites in tourist areas, I have been complicit in forming. On my previous visit in this capacity, the local expert on the rock art, a non-Aboriginal expert, assured me there were no Aborigines for this place left. What he meant was there were no Aborigines for this place, in this place, or in the 'form' he would accept as authentic. This local expert saw himself as the gatekeeper of this place, the one who had saved it from the ravages of tourism, the one who had collected up artefacts and bodies and kept them for safe keeping, the one who, as a National Parks Wildlife Service officer, has regulated its recent development. For this expert, who at this time had his own private store of Aboriginal bodies and arte-facts in a shed in the backyard of his Carnarvon Gorge home, it was perhaps more comfortable to imagine the Carnarvons with no indigenous descendants.

He felt he knew more about this place than most, including any Aboriginal descendants who might lay claim to this area or trace lineage back to this place from their distant homes throughout Queensland. He probably does—if one measures knowledge as possession of place-based artefacts and information.

Jackie

Displacement—During the late 1920s Rita, my mother, and her family were rounded up by the troopers and sent on the back of a cattle truck to the then Barambah Aboriginal reserve, later known as Cherbourg. Rita reminisces about the golden age of traditional Aboriginal existence in a manner which most people would find beautifully interesting—glimpses of a pristine past that was a common picture prior to the white invasion. She states:

> My mother would make soap from the leaves of a tree and unfortunately for us there was no excuse not to take a bogey. Goanna fat was used for cuts and scratches on bare feet and limbs as well as soothing treatment for aches and pains; eucalyptus leaves for coughs and bark for rashes and open wounds; witchetty grubs for babies' teething while charcoal was used for cleaning teeth. Bush tucker also thrived in this environment, and we were never left with empty bellies. The men would go hunting for kangaroos, goannas, lizards, snakes, porcupines with their spears, boomerangs and nulla-nullas while the women gathered berries, grubs, yams, edible roots, wild plums, honey and waterlilies with digging sticks. Children always accompanied the women as there was less likelihood that animals would run away when disturbed. The creeks supplied an abundant and rich source of fish—jew, yellow belly, perches and eels.

While my mother recalls those idyllic days of Carnarvon Gorge it belies a history of an opposite kind, that is, the bloody massacres which occurred not only here but which followed a typical pattern right across the continent. Many massacres were ritualised to demonstrate the white superiority and power that was being shaped on the frontier at the time.

The perpetrators were seen as the pioneers and heroes who had entered into the place they called the 'wilderness' but which we called home. The quest of these men was to tame the natives and the land. Their mystique and greed cost my people dearly. There was only one thing they could do with us—get rid of us completely so they wouldn't have anybody else to consider—because then our tranquil home would become their fortress and recreational areas through force and theft.

Conquerors, no matter how they got what they wanted, were then issued with licences to kill in order to protect their newly acquired property. These licences came in many different disguises; be they poisoning flour or waterholes, burying Blackfellas alive in the sand, tying them around trees to use for shooting practice, all this and much more.

The scarcity of white women in colonial times meant that the colonisers exploited Aboriginal women. The result was rape and 'half-caste' children whom their biological, white fathers certainly did not want or own. Other factors and atrocities which contributed to the retaliation was the killing of Aboriginal dogs and game, the dispossession of hunting grounds through invasion, and the exclusion and destruction of sacred sites.

The Jiman who also bordered Carnarvon Gorge had enough of the harsh treatment of their women at the hands of the invaders. After killing all present at the Fraser homestead their revenge acted as the catalyst for a six months 'little war' which was waged against them by white vigilantes. This vendetta was carried out through the Dawson and Burnett districts which number almost a thousand square kilometres. As punishment for killing the Fraser family, all Aboriginal people—men, women and children—in the area were shot down as they ran. It had all become an uncontrollable rage and wholesale slaughter of innocent human lives. No effort was ever made to bring the white murderers to court. The killing and mutual hatred went on long after, and all over Australia.

The people of my mother's generation display a profound lack of bitterness about their lot, something which I find frustrating and amazing. It is probably this trait though, this lack of bitterness, that has polarised and politicised so many of my generation to become active about the continuing injustices occurring to Aboriginal people.

We were once the proud custodians of our nation and now our way of life would be controlled and irreversibly shattered by colonialism. Aboriginal people could not choose how or where to live again, they were told to stay in one place and dictated to by government officials. Once the permit system allowed Aboriginal people freedom from the reserves, it was we who became the indigenous tourists of this country.

Jane
Forever Frontier—In this part of Queensland the violence is ever-present in the absence of Aborigines. Here the Aboriginality of place is confined to rock art which lines the walls of caves. Early tourist brochures on the gorge spoke of 'the old dead Myall race'. Contemporary brochures evade the history of violence, stating 'the artists are gone', and alluding to the further 'tragedy' of losing the 'heritage they left' (Queensland National Parks and Wildlife Service 1991).

In the Carnarvons, the violence of colonisation has persisted through the transitory but continual possession of this place by tourists. In 1993, some 400 tourists each weekend confirm this is no longer Bidgera country, but a public space, a national space. Non-Aboriginal tourists are clearly unsettled when they confront indigenous tourists walking along the trail, even more so when Aunty Rita strikes up a conversation and announces 'I was born here'. This is not to say the non-Aboriginal tourist does not expect or want to meet Aboriginal people when out bush, most do; it satisfies their desires for an authentic Australian experience. And many tourists in Central and Northern Australia have these desires well met in parks under joint management or in tourist ventures run by Aboriginal communities. But in Carnarvon Gorge the Huggins family seems 'out of place', their claims of ownership, along with their presence in the Gorge itself, are rendered unbelievable by both the mythology of a place thoroughly cleared of Aborigines and the iterative practice of tourist discovery.

The National Parks sign we confronted on our first walk through the Gorge together directs the tourist to gaze upon an imaginary Aboriginal time, in what is now an ambiguously Aboriginal place. If I turn the sign around I can gaze upon the imaginative projections which have ensured this place was possessed as a tourist space, serving the transient in search of the authentic.

In 1937, Danny O'Brien of the Royal Geographical Society of Australasia (Queensland Branch) led the first of a number of expeditions to the Carnarvon Ranges. These were expeditions of a modern Australia, a modernising Australia. The group travelled not by horseback, but by train and car. They did not harbour the reluctance for photography noted in Paul Carter's accounts of the early explorers (Carter 1992, p. 31). These latterday geographical expeditions included photographers and film-makers. Trails were not carved through unknown lands, the groups followed local guides from nearby pastoral leases and performed ritualistic ceremonies of governmentality through various civic welcomes and farewells. By this stage in the settlement and development of non-Aboriginal Australia, this place at best could be described as 'little known land' (O'Brien 1942, p. 8). The area was already surveyed and mapped, no frontier cartography was necessary. Less than ten years after the first expedition, the Carnarvons had been captured by an areal photograph run. Little remained to be claimed by this expedition. All major natural features were already named, even renamed: 'We stood on Consuelo Tableland...and viewed the great valley of Lethbridges' Pocket with its deep walls...We saw in the distance, Tyson's Nugget, Mount Lethbridge (also known as Kenifps Lookout), Vandyke Creek, veering away towards Springsure and Mount Sugarloaf' (O'Brien 1950, p. 2).

All that was left for this mimic explorer to name were modest geographical features or newly constructed engineering works. He named after himself both a road he lobbied local authorities to grade and a makeshift causeway one of the expeditions made.

O'Brien's writings on his expeditions had the effect of imaginatively depopulating this place. There is the familiar decanting of Aboriginal occupants or, in this case, the memory of Aboriginal occupation. But O'Brien also evacuates non-Aboriginal settlers. He reaches Carnarvon Gorge through technology, engineering and local knowledge, yet he depicts the place as untouched by settlement—where 'the footprints of the white man are still missing from many of its most secluded places' (O'Brien n.d.). In depopulating the place, O'Brien could label his first trip to the Gorge a 'pioneer expedition', and speak convincingly of the 'insurmountable difficulties' they faced in traversing 'new country'. O'Brien was beset with a nostalgia for an earlier age of masculinist exploration when geographical knowledges of description and mapping were necessary for the possession of place. His 'undiscovered' Carnarvons were surrounded by pastoral leases, and by 1932 the Gorge itself had been designated by the state authorities as National Park. His desire to reinvent the masculinist project of pioneering exploration was most seriously challenged by the presence of a small travellers' cottage at the mouth of the Gorge, built before the first expedition by the local Country Women's Association. O'Brien's writings empty the land of people and reinvent the frontier as an uninhabited space, a virginal space for his taking. This imaginative manoeuvre still resonates through the tourism industry in Australia, and particularly in new forms of cultural tourism and ecotourism.

O'Brien and the Royal Geographical Society (RGS) attempted to consolidate the dubious masculinity of their expeditions through recourse to science and technology. The very description 'expedition', from the noun 'expedite', evokes an authorised, systematic venture with an intent of clearing difficulties, or hastening progress. There is little evidence of such systematic practices. Instead O'Brien orchestrated certain technological rituals, making the first chartered air flight over the Gorge and making the first wireless transmission from the Gorge. What was written about the early expeditions reads as travelogue, and the intent was to establish this place as 'scenic' in the wider imagination of Queenslanders.

O'Brien's many expeditions to this area during the late 1930s, 1940s and into the 1950s, were driven by his desire to see the Carnarvons become one of Australia's key tourist destinations. The 'pioneer expedition' was considered triumphant because it confirmed the Gorge could be reached by car over level country, and that, if an all-weather road was built, the Gorge could be reached in a 'comfortable' two day motor trip from Brisbane. O'Brien took his lantern slide lectures on the Gorge throughout central and southern Queensland. He was, by his own description, 'the honorary advocate of Carnarvon range development', and his rewards were 'the knowledge that there now exists a tourist traffic to Carnarvon National Park' (O'Brien 1942, p. 36). It was an advocacy of some interest to the Queensland government, which had already established a Tourism Bureau and was interested enough in the Gorge to supply a Director of Films and Cinematographer for the 'pioneer expedition'.

The RGS expeditions became increasingly ones of pleasure for the groups of twenty or so. By 1956 a circular letter encouraging people to take a Royal Geographical Society trip to the Gorge was still referring to the ventures as 'expeditions'. But the title 'expedition' was becoming increasingly irrelevant and was giving way to other notions:

> An outstanding feature of recent expeditions is that, though we are not now attracting so many scientists, we are getting a great many intelligent younger people. Thus the real aims of our Society are being accomplished, and instead of geology, anthropology, etc., just being laborious subjects in the lecture room, they are now 'picnic' subjects—education by pleasure. Our students don't get nervous collapses caused by study, though some have collapses caused by Cupid...This is Queensland's biggest picnic trip (Letter, RGSA (Q) 1956).

But, despite some interest from the Queensland Tourism Bureau in the area, early evaluations of its tourist potential were disappointing for O'Brien. Tourism officials and the press who visited the area in 1949 suggested the area was 'overrated as a tourist attraction', had 'disappointing' scenery and poor accessibility. At best, the area might serve as a low cost holiday destination for 'low income western residents' (*Charleville Times* 22 December 1949).

Carnarvon Gorge has remained a relatively inaccessible place, a destination only for the keen or the local. Tourist development remains modest. In 1999 the main road to the Gorge is still unsurfaced. There is no Yulara, or Alligator Hotel, simply the log cabins of Oasis Lodge and the National Parks camp ground. The pace of recent tourist development of the area has been regulated by a local non-Aborigine, a self-appointed gatekeeper of Carnarvon. O'Brien saw the development of the Gorge as depending on the intervention of engineering, the construction of roads, and the building of resorts. This more recent gatekeeper has used engineering and technology for different ends. In the early 1980s, military helicopters airlifted timber into the Gorge so that massive boardwalks, the first of their kind at art sites in Australia, might be built. The boardwalks were designed to keep tourists away from the art, they protect 'heritage'. But the boardwalks also opened the art to more precise and regulated scrutiny, directing where visitors go, drawing the visitor's attention to specific details, interpreting the art in terms of technique and meaning, and presenting the signature of 'heritage' on the landscape of Carnarvon Gorge. Structures built to protect the art, inscribe ongoing possession of this place by tourism.

Jackie
Reclaiming Place—The High Court Mabo decision attempts to rectify some of these injustices in the future through the land claim process. Coincidentally the claim made on behalf of the Bidjara people in June 1993 is the largest in Queensland. Five other clan groups may also lay claim to it due to its usage by them for ceremonial purposes.

Tensions are mounting between the Aboriginal traditional owners and interest groups, pastoralists, the National Parks and Wildlife Service, and, of course, the white experts. Decisions will reflect all parties' concerns and will be facilitated though a three member Aboriginal Land Claims Tribunal which, incidentally, has no Aboriginal membership. Negotiations with the Aboriginal owners will depend upon the information so evidently tainted by dispossession and the memory of violence.

For this reason Rita and others will be pressured to remember the childhood memories of her place, when once they were forced to deny and forget them through the protection and assimilation policies of the day. How ironic it is that now every detail to recall and articulate forms the basis for claims, and is thus honourably encouraged and documented. History finds a way of reinventing itself, particularly for Aboriginal people, for better or for worse. At this early stage it would be ludicrous to suggest that all parties will be completely satisfied with a desirable outcome.

In conclusion, I want to return to the early tourists. The desire for tourist pleasure was shadowed by the memory of violence. Mrs Hamlyn-Harris was one of the first women to go on a Royal Geographical Society Carnarvon expedition. She wrote a series of poems about this place as a tribute to the intrepid leader Danny O'Brien. In one of her poems, simply entitled 'Carnarvon Gorge', she contemplated the history that led to her encounter with this place.

> Yes, what have we done?
> The past is dead by our own hand. The olden race is gone.
> They are but ghosts, those spirits of the wind and trees
> That sigh such doleful threnodies!
> We of To-day by duty bound must wake

And silence them with Joy's sweet laughter ringing,
And children's voices singing! We must arise, and of this primal
    playground make
A resting place for toiling families of the drought stricken west
Needful of Nature's healing and of rest.
Thus shall we at last
Make humble expiation for the past (Hamlyn-Harris, n.d.).

Turning Kooramindanjie Place into Queensland's 'biggest picnic ground' has not made 'humble expiation' for the past.

6

The Future for Feminist Geography

Jackie Huggins's and Jane Jacobs's assessment of Kooramindanjie Place/Carnarvon Gorge raises a host of issues about academic research, especially how geographers and historians position themselves as observers and writers. Jackie Huggins is a historian and Aboriginal woman whose mother accompanied her on this trip to their ancestral lands. Jane Jacobs, as a geographer writing in the light of the postcolonial critique, engages with a dual challenge, positioning herself within this Queensland landscape as white 'expert' in Aboriginal heritage, rock art and tourism management, and as a postcolonial feminist geographer, to take part in a dialogue with Jackie Huggins. The result is a multifaceted story. Jane, as an academic, presents the white history of heroic settlement and development as a tourist destination replete with 'lost' Aboriginal heritage while she also grapples with the identity and representational issues raised by postcolonialism. She converses with Jackie as an Aboriginal woman historian whose family history is written in this place of ongoing colonisation. This history ranges from first contact, to ongoing violent dispossession and land claim. The conversation between Jackie, Jane and Auntie Rita is complex as they negotiated their reflections and absorbed their different connections with this place. This is an example of the working out in one location of aspects of a multifaceted postcolonial geography.

If colonisation is a military invasion and takeover of one country by another, it is also an ongoing process of economic, political and cultural occupancy. This occupancy involves the transformation of identities for those on both sides of the racial and military divide. Some see Australia as a 'settler society'. Such a benign label ignores the violent nature of occupancy, and it fails to anticipate the diverse reaction to the political separation from the United Kingdom. Against this view, postcolonialism rewrites Australian history as an ongoing imperialist project; it admits the voices of those silenced, marginalised and incarcerated by that history, and it includes a politics of resistance, disruption, visibility, difference and positionality by those of Aboriginal, Torres Strait Islander and non-English speaking background. Feminists suggest that the colonial project is also a gendered one; however, postcolonial feminist geographers have emphasised the continuing processes of colonisation—by travellers and geographers—and the contribution of race and ethnicity in constituting landscapes, neighbourhoods and workplaces. The postcolonial perspective has moved other dynamics onto centre-stage, so that gender is no longer the key concept, the taken-for-granted determinant or definer of oppression, identity or difference. Gender becomes one of a number of power-soaked relations and dimensions inhabited by one person at any time and place.

In this shift, from a focus on gender to a range of other dimensions which create an increasingly fluid identity, feminist geography has drawn on postmodern theory. Its context is post-1970s capitalism. This is a capitalism whose Western economic hegemony is challenged by newly industrialising Asian countries; it is organised by multinational finance capital within a regime of flexible accumulation, and characterised by a new style of architectural design and adornment. Postmodern discourse links cultural and economic changes to a crisis of representation; this emanates from feminists, blacks, gays and other marginal groups who have raised profound questions about the way in which cities, individuals and societies have been conceptualised. Yet, interestingly, the postmodern feminist is akin to the earliest feminist; like her earlier counterpart, the postmodern feminist finds bodies of supposedly universalist knowledge and their assumptions wanting in their applicability to women. These knowledges may take the form of a neoclassical assessment of the urban environment or a postmodern homage to Los Angeles, but what postmodern feminist geographers argue, is that such work denies the existence of women as a group, and that it ignores the body of critical feminist discourse which makes sense of that fractured and differentiated world, whether it is a city or shopping centre.

Postcolonial and postmodern feminist geographers emphasise fractured and multifaceted identities in place. These identities may be differentiated by race, ethnicity, sexuality or location within a postcolonial set of power dynamics, but the emphasis involves a dramatic transformation in the theorising of women which first informed feminist geography. In its initial formulation, feminist geography in Australia, the United Kingdom, New Zealand and North America was heavily influenced by Liberal, Radical and Socialist Feminist thinking. This thinking suggested that women were a discrete, identifiable group, despite some internal differentiation. These feminisms were inspired by the unease which first infused the women's liberation movement: anger at the oppressed position of women, anger at the men who benefited from this situation, and optimism at the possibilities for change within or outside the state. Such views had a number of profound implications. In the case of radical feminist geography, the idea of women as a unified group distinguished by their difference from, and oppression by, men led to a view of cities divided by gender, and to an activist political agenda to change this situation. Analyses documenting 'Cities of fear' revealed how women were being kept in a state of fear, immobilised and constrained in their activities in urban environments by the threat and actuality of male violence. The suggested solutions ranged from urban redesign and lighting to a curfew on men!

Similarly, if less obviously, studies by liberal feminist geographers, which mapped inequality in the geography profession, and across cities or regions, illustrated gross divides between women and men in income, educational achievement, political representation and seniority in employment. These cartographic exercises led to policies which attempted to address the barriers to equal participation and challenge women's position in low level jobs, in the home and outside it, in the public sphere. So too, socialist feminist geography saw women as a separate group, with a relationship to capitalism which was driven primarily by class location and shaped by gender. Thus it was that capitalism and patriarchy were related, both theoretically and empirically, either as a unified system or a dual system where capitalism or patriarchy was dominant. Either way, women and men were divided and further differentiated primarily by class, though what remained fundamental in this perspective was the oppression of women within systems of production and social reproduction; these systems then became sites of ameliorative action.

Such then was the fundamental assumption of feminist geography: that women were a group united by their inequality and oppression by men. They may have been further defined and differentiated by their class, ethnicity or sexuality; but primarily their location in space and place was the result of their sex and their subordinate relationship to men. They were in every sense of the word Placebound: bound to their place in society by a range of assumptions and social relations which, while historically variable, and open to challenge and amelioration, were remarkably immutable.

This has been the historically dominant view in feminist geography. I see this view as present in the early 1970s, surviving into the early 1990s, and yielding to the postcolonial–postmodern view of women as a category defined by a range of discourses which vary across time and space. Women in contemporary feminist geography are therefore Placebound in a different sense of the term. For they are now going somewhere, both conceptually and actually, towards a place which has yet to be defined and which they may never reach. This is the geographical equivalent to Derrida's continuous deferral of meaning, where any fixing in words or space of a word 'woman' is but a temporary event. Women are forever bound for a different place, one where there continues to be a range of dominant and lesser discourses jostling for definitional dominance, and where the outcome cannot be predicted. Metaphorically, women are in the driving seat in this future, placebound journey.

Women may now be defining themselves in space, but there are losses in this vision. The new vision dissolves the anger and the unified voice of a 'women's movement' which focused on women as a group challenging men. The focus is now on texts rather than women, discourses rather than oppression, and on the present and its multifaceted complexity rather than a collective future. Such a view allows diversity, multiple identities and a range of oppositional discourses to bloom, so that, for instance, the complicit and central role of women in the imperial project can now be made apparent. Yet women remain, in many ways, bound to places not of their own choosing or making. The challenge is how to reconcile the two senses of placebound, so that women can make a place of their own.

■

Glossary

appropriation—to take possession of. Such an action may or may not be legitimate or sanctioned. For example, within the Marxist framework, the products of a worker's labour is appropriated by the capitalist who then returns only some of its value in the form of a wage. In this situation appropriation is unjust.

constitutive—to actively create. For example, a set of constitutive social relations are ones which are created and negotiatied by those involved.

contingent—to be dependent on context and historical location for meaning. Also to be unique, individual and lived. A contingent proposition may be true and may equally be false; the matter is contingent or dependent on factors external to the proposition itself.

destabilise—to render unsure or uncertain. To take off-balance.

discourse—in Linguistics 'discourse' refers to a stretch of language larger than a sentence. Its analysis involves looking at semantics, style and syntax and the sequence of sentences as well as their structure. Building on such a notion, Michel Foucault saw in discourse the means by which language was connected to modes of thought, individual subjectivity, social institutions and structures. Each utterance, or discourse, involves a statement of a subjective position in language, but the speaker or writer also has a social position and therefore some social power that shapes the status and impact of the utterance. All these elements—what is said, by whom, from what social position and with what effect—forms a discourse.

discursive—discursive fields consist of competing ways of giving meaning to the world and of organising social institutions and processes. Social structures and processes are organised through institutions and practices such as the law, politics, religion, education and the media, each of which is located within a particular discursive field. Within each field an individual will have a discursive position, that is a set of ideas and beliefs which are uttered and assume certain meaning and power as a result of that social position. As a consequence differential discursive power relations will be formed between competing beliefs and ideas depending on the social power wielded by whomever is making the utterance. A High Court judge rendering a legal opinion will not only have a powerful discursive position but will be near the apex of a set of legal discursive power relations which in turn put the lay person in a relatively powerless position and the trained lawyer in an intermediate position.

double coding—a term used by the architectural critic Charles Jencks to describe the tendency of postmodern builders to draw from and use a range of eras, traditions and localities when designing their structures. The result is that any one building will not have one style but rather will have at least two, be it distinctive elements of the international modern style combined with ancient Gothic or Chesterfield or any other combination, to double code a structure. This double coding is obvious and, unlike eclecticism, involves a clash and discordance between the styles or design elements used in any one design.

essentialist—a notion that particular things have essences which serve to identify and make them distinctive and which are critical to their meaning and being. The term has been used by some feminists to question a tendency to define women in terms of their reproductive capacities or anatomy. This is seen as essentialist and usually dangerously so, confining women as much as rigidly defining them by their biology.

foregrounding—to bring to the fore; to make obvious and clear.

gaze—the way someone looks to thereby give the person or thing looked at some meaning. Feminists have argued that the gaze in film especially is masculine; that through a range of technical mechanisms to do with lighting, camera angles and the construction of image sequences the male gaze, or the look from the male actors, is identical to that of the camera. The film, and by implication a range of other image-dependent media, is therefore a product of and reproduces a dominant, male way of looking, one which tends to objectify women.

gendered bodies in space—if women and men are not defined and shaped by their biology, then they are gendered, that is they acquire their identities as women and men through social processes. As they move into and through space, this gendered identity will mark and give a certain character to that space while also affecting those who are in the space. For example, the football dressing room after a match is very much a male space with its exposure of male bodies, dirty and sweaty from the match. But such a space also accommodates a culture with men as part of a team defined by physical and technical prowess, competitiveness and an active rejection of 'feminine' values of intimacy, gentleness and compassion. The men, their bodies and that change room assume a character as a result of that combination of social markers and conventions.

grounded—to bring down to earth; to illustrate through concrete example. To make well-founded on a firm foundation.

iterative practice—a way of speaking, arguing or acting which consciously repeats, incrementally adds to and builds on what went before. A process that builds a position in a cumulative way.

necessary—logically essential for any concept to have meaning: true, absolute, basic and general notions above history and context. That there are four apples in this bowl is contingent; that there are four apples in a bowl containing two pairs of apples is necessary, that is required by the meaning of the key concepts.

occludes—obscures and thereby excludes.

post-Fordist—the end point after the transition from a modern or industrial to a postmodern or post-industrial society in which the means of producing goods has changed significantly. The model of the industrial form of production is the Ford Motor Company plant in Detroit in the early twentieth century. Here, large numbers of workers; mostly unionised, well-paid and relatively unskilled men, laboured on a production line doing the same minor work task throughout the day to produce large numbers of standardised products for a mass market. This Fordist model of production is now seen by many commentators to have been replaced by a post-Fordist model. In this the production line is replaced by small-scale batch production by mixed groups of highly skilled, non-unionised workers organised in flexible teams. Many of the components are produced off-site by sub-contracted firms and arrive just in time for their inclusion. In this post-Fordist era, the products are customised and carefully targeted to fit into niche markets.

problematise—to raise questions about or to see as a problem rather than take something for granted.

racialised—to construct someone in racial terms. A racialised workforce is one that is defined in terms of its racial composition.

restructuring—any society or economy has a structure which can be conceptualised and theorised in a particular way. When dealing with the Australian economy over the last thirty years, one of the most profound changes has been its restructuring away from an emphasis on manufacturing and industry towards the service sectors. This has been impelled by a fall in the profitability of industrial production such that economic restructuring can be seen as the movement of capital and labour from one sector to another in an effort to restore profit rates to an acceptable level.

round of accumulation—in the Marxist framework, capital has to be invested in some form of production. Human labour is applied to raw materials and, within some form of production system, a range of goods and services are produced that contain more value than the component parts. For their value to be realised these goods must enter and be purchased in the market. The circuit of investment, production and consumption is a round of accumulation. At the end of it, more value exists as a result of the application of human labour. Capitalism is a system which is expansive—each round of accumulation will be larger than the preceding one unless a crisis occurs at any point in the chain. Each round of accumulation can be associated with a particular production site and landscape; for example, rounds of industrial accumulation produce very different landscapes from those related to mining or trade.

sexed bodies in space—some contemporary feminist theory argues that the nature of bodies is indeed fundamental to identity. While these bodies are still seen as inscribed and given meaning by a range of social discourses, they are also corporeal and real. Sexed bodies are therefore seen as the result of an interaction between biological matter and social constructions. How such sexed bodies then move into and shape space or are shaped by spaces involves recognising the importance of physicality to that process. A gymnasium is one space in which women's and men's bodies are made over, stretched and reshaped in ways that affirms their physical character.

■

Further Reading

Chapter 1 Liberal Feminist Geography

The 'Geography of Women' literature includes some useful compilations about women's location, activity patterns and discrimination; for instance, *Women and Space: Ground Rules and Social Maps* by S. Ardener (1981), St Martin's Press, New York; *Her Space Her Place: A Geography of Women* by M.E. Mazey and D.R. Lee (1983), Association of American Geographers, Resource Publications in Geography, Washington, DC; *Gendered Spaces* by D. Spain (1992), University of North Carolina Press, Chapel Hill. There is a host of specific studies on women and transport, migration, development and ageing too numerous to mention here. Good reviews of this literature are provided in 'On Not Excluding Half of the Human in Human Geography', *The Professional Geographer*, vol. 34, no. 1, pp. 11–23, by J. Monk and S. Hanson (1982) and 'Women and Geography: A Review and Prospectus', *Progress in Human Geography*, vol. 6, pp. 317–66 by W. Zelinsky, J. Monk and S. Hanson (1982), as well as in *Women and Geography Study Group of the IBG* (1984), Geography and Gender, Hutchinson, London.

Chapter 2 Socialist Feminist Geography

One of the earliest and clearest expositions of the socialist feminist position is Juliet Mitchell's 1966 essay which became the book *Women's Estate*, Vintage Books, New York, 1971. For an overview of socialist feminist thinking see 'Socialist Feminisms' by Johnson in S. Gunew (ed.), *Feminist Knowledge: Critique and Construct*, Routledge, London, 1990, pp. 304–32.

Some of the best socialist feminist geographies are: 'Towards an Understanding of the Gender Division of Urban Space', Linda McDowell's essay in *Environment and Planning D: Society and Space* vol. 1, 1983, pp. 59–72; 'Industrial Change, Domestic Economy and Home Life' by S. Mackenzie and D. Rose in J. Anderson, S. Duncan and R. Hudson (eds.)

Redundant Spaces? Studies in Industrial Decline and Social Change, Academic Press, London, 1983, pp. 155–200 and *Spatial Divisions of Labour: Social Structures and the Geography of Production* by D. Massey, London, Macmillan, 1984.

The position of this discourse in the late 1990s is best indicated by the work of Katherine Gibson and Julie Graham, especially J. K. Gibson-Graham, *The End of Capitalism (As We Knew It)*, Blackwell, Oxford, 1996.

For work which further details the Geelong case study, see Louise C. Johnson (1990b), 'New Patriarchal Economies in the Australian Textile Industry', *Antipode*, vol. 22, issue 1, pp. 1–32; and Johnson 'Restructuring and Socio-Economic Polarisation in a Regional Industrial Centre', in K. Gibson et al. (eds.) *Restructuring Difference: Social Polarisation and The City*, Australian Housing Urban Research Institute, Working Paper No. 6, Melbourne, 1996, pp. 43–57.

Chapter 3 Radical Feminist Geography

To gain a quick sense of the essential nature of radical feminist thought two of the classic texts are Shulamith Firestone's *Dialectic of Sex*, The Women's Press, London, 1979, and Kate Millett's *Sexual Politics*, Virago, London, 1977. A good, critical overview of the position is Hester Eisenstein's *Contemporary Feminist Thought*, Unwin, London and Sydney, 1984; while a recent collection by Australian feminists is Diane Bell and Renate Klein's *Radically Speaking*, Spinifex, Melbourne, 1996 is a contemporary statement of the radical feminist position.

Radical feminist geography is best represented by the 'patriarchy debate' with Jo Foord and Nicky Gregson's 'Patriarchy: Towards a Reconceptualisation', *Antipode* vol. 18, issue 2, 1986, pp. 186–211, and subsequent comments in *Antipode* vol. 18, issue 3, and vol. 19, issue 2. In addition, there is the 'Geography of Fear' literature, spearheaded by Gillian Valentine's 'The Geography of Women's Fear', *Area* vol. 21, issue 4, 1989, pp. 385–90, and 'Women's Fear and the Design of Public Space', *Built Environment* vol. 16, issue 4, 1990, pp. 288–303.

Chapter 4 Postmodern Feminist Geographies

One of the first and most important essays on the postmodern was Frederic Jameson's 'Postmodernism: The Cultural Logic of Late Capitalism' *New Left Review*, vol. 146, 1984. In geography, the postmodern has been most eloquently explored by David Harvey in *The Condition of Postmodernity*, Blackwell, Oxford, 1989; Mike Davis in *City of Quartz*, Vintage, London, 1990; and Edward Soja in *Postmodern Geographies*, Verso, London, 1989 and *Thirdspace*, Blackwell, Cambridge, Mass., 1996.

This work has also been the subject of stringent critiques from feminist geographers, especially by Doreen Massey in 'Flexible Sexism', *Society and Space* vol. 9, 1991 and Steve Pile and Gillian Rose in 'All or Nothing', *Society and Space*, vol. 10, 1992. The most detailed postmodern analysis of contemporary geography is Gillian Rose's *Feminism and Geography*, Polity Press, London, 1993. The best recent study of identity formation, which focuses primarily on class and ethnicity, is Miller et al.'s *Shopping, Place and Identity*, Routledge, London, 1998.

Chapter 5 Postcolonial Feminist Geographies

The best way to grasp the breadth and diversity of postcolonialism is to consult some of the excellent collections of key writings, such as P. Mongia's *Postcolonial Theory. A Reader*, Arnold, London, 1996.

Some of the key works are *Nation and Narration* by Homi Bhabha, Routledge, London, 1990; *The Location of Culture*, Routledge, London and New York, 1994; 'Culture, Community, Nation' by Stuart Hall in *Cultural Studies* vol. 7, issue 3, 1993, pp. 349–63; and *In Other Worlds: Essays in Cultural Politics* by Guatri Spivak Methuen, London and New York, 1987.

Some Australian studies with a postcolonial framework include Tony Birch's 'A Land So Inviting and Still Without Inhabitants' and 'Erasing Koori Culture from (Post-)colonial Landscapes', in *Text, Theory, Space. Land Literature and History in South Africa and Australia* by Kate Darian-Smith, Liz Gunner, and Sarah Nuttall (eds.), Routledge, London and New York, 1996, pp. 173–85 and Louise Johnson 'Occupying the Suburban Frontier: Accommodating Difference on Melbourne's Urban Fringe', in *Writing Women and Space: Colonial and Postcolonial Geographies* by A. Blunt and Gillian Rose (eds.), 1994, Guildford Press, New York, 1990, pp. 141–68.

Postcolonial geographies include *Geography and Empire* by A. Godlewska and Smith (eds.) N. Blackwell, Oxford, 1994; Felix Driver's 'Geography's Empire: Histories of Geographical Knowledge', *Environment and Planning*, vol. 10, 1992, pp. 23–40; Jane Jacobs' *Edge of Empire*, Routledge, London and New York, 1996 and Doreen Massey's *Space, Place and Gender*, Polity, Cambridge, 1994.

The most comprehensive postcolonial feminist geography collection is A. Blunt and Gillian Rose's *Writing, Women and Space: Colonial and Postcolonial Geographies*, The Guildford Press, New York, 1994. See also Mona Domosh's 'Towards a Feminist Historiography of Geography', in *Transactions*, Institute of British Geographers NS, vol. 16, 1991, pp. 95–104.

References

Australian Bureau of Statistics (1971; 1976; 1981; 1986; 1996) Census of Population and Housing. Catalogue No. 2020.0 Time Series Community Profile, Australian Government Publishing Service, Canberra.

Australian Bureau of Statistics (1976; 1986) *Census of Population and Housing Small Area Data*, Australian Government Publishing Service, Canberra.

Australian Bureau of Statistics (1992; 1997) *Women in Australia* Catalogue No. 4113.0, ABS, Canberra.

Australian Bureau of Statistics (1982; 1995; 1997; 1999) *Year Book Australia*, ABS Catalogue No. 1301.0, Canberra.

Australian Bureau of Statistics (1997) *How Australians Use Their Time*, Catalogue No. 4153.0, Australian Government Publishing Service, Canberra.

Adlam, Diana (1979) 'The Case Against Capitalist Patriarchy', *m/f* 3, 83–102.

Adler, Sy and Brenner, Joanna (1992) 'Gender and space: lesbians and gay men in the city', *International Journal of Urban and Regional Research*, 16 (1), 27.

Ahmad, Aijaz (1995) 'The politics of literary postcoloniality', *Race and Class*, 36 (3), 1–20.

Aldrich, Robert and Wotherspoon Gary (eds), (1992). *Gay Perspectives: Essays in Australian Gay Culture*, Department of Economic History, Sydney University, Sydney.

Allport, Carolyn (1986) 'Women and suburban housing: post-war planning in Sydney, 1943–61' in *Urban Planning in Australia: Critical Readings* (eds J.B. McLoughlin and M. Huxley), 233–250. Longman Cheshire, Melbourne.

Allport, Carolyn (1987) 'Castles of Security: The New South Wales housing commission and home ownership, 1941–1961' in *Sydney: City of Suburbs* (ed. M. Kelly), NSW University Press in association with the Sydney History Group, Sydney.

Amos, V. and Parmar, P. (1984) 'Challenging imperial feminism', *Feminist Review*, vol. 17, 3–20.

Anderson, J., Duncan, S. and Hudson, R. (eds), (1983). *Redundant Spaces in Cities and Regions? Studies in Industrial Decline and Social Change*, Academic Press, London.

Anderson, Kay (1990) 'Chinatown re-oriented: a critical analysis of recent redevelopment schemes in a Melbourne and Sydney enclave', *Australian Geographical Studies* 28 (2), 137–213.

Anderson, Kay (1991) *Vancouver's Chinatown: Racial Discourse in Canada, 1875–1980*, McGill-Queens University Press, Montreal.

Ang, Ien (1995) 'I'm a feminist but... "other" women and postnational feminism' in *Transitions. New Australian Feminisms* (eds B. Caine and R. Pringle), 57–73, Allen & Unwin, St Leonards.

Anthias, Floya (1980) 'Women and the reserve army of labour: a critique of Veronica Beechey', *Capital and Class* 10, 50–63.

Anthias, F. and Yuval-Davis (1983) 'Contextualizing feminism: gender, ethnic and class divisions', *Feminist Review* 15, 62–75.

Anzaldua, Gloria (1987) *Borderlands/La Frontera. The New Mestiza*, Spinsters, San Francisco.

Appadurai, A. (1996) *Modernity at Large: Cultural Dimensions of Globalization*, University of Minnesota Press, Minneapolis.

Appiah, Kwame Anthony (1996) 'Is the post- in postmodernism the post- in postcolonial? in *Postcolonial Theory. A Reader* (ed P. Mongia), 55–7, Arnold, London.

Ardener, Shirley (ed.) (1981) *Women and Space: Ground Rules and Social Maps*, St Martin's Press, New York.

Attwood, B. (1989) *The Making of the Aborigines*, Allen & Unwin, Sydney.

Australian Housing Bulletin (1947) *Government Sponsored Housing in Victoria* No. 13, Housing Division, Department of Works and Housing.

Awatere, Donna (1984) *Maori Sovereignty*, Broadsheet, Auckland.

Badcock, Blair (1997) 'Recently observed polarising tendencies and Australian cities', *Australian Geographical Studies* 35 (3), 243–59.

Bagguley, P., Mark-Lawson, J., Shapiro, D., Urry, J., Walby, S. and Warde, A. (1990) *Restructuring: Place, Class and Gender*, Sage, London.

Bandler, Faith (1989) *Turning the Tide*, Australian Institute of Aboriginal Studies, Canberra.

Barrett, M. and McIntosh, M. (1982) *The Anti-Social Family*, Verso, London.

Barrett, M. and McIntosh, M. (1985) 'Ethnocentrism and socialist feminist theory', *Feminist Review* 20, 24–47.

Barry, K. (1979) *Female Sexual Slavery*, New York University Press, New York.

Barry, K. (1995) *The Prostitution of Sexuality: The Global Exploitation of Women*, New York University Press, London.

Baudrillard, Jean (1983) 'The Ecstasy of Communication' in *Postmodern Culture* (ed. H. Foster), Pluto, London, 126–34.

Baum, S. and Hassan, R. (1993) 'Economic restructuring and spatial equity: a case study of Adelaide', *Australian and New Zealand Journal of Sociology*, 29 (2), 151–172.

Beckwith, Karen (1977) 'The status of women professionals in geography', *Professional Geographer*, 29 (4), 404–5.

Beddington, N. (1991) *Shopping Centres: Retail Development, Design and Management*, Butterworth Arch, Oxford.

Beechey, Veronica (1977) 'Some notes on female wage labour in capitalist production', *Capital and Class*, 3, 45–66.

Bell, David (1973) *The Coming of Post-industrial Society*, Penguin, Harmondsworth.

Bell, Diane (1983) *Daughters of the Dreaming*, McPhee Gribble / George Allen & Unwin, Sydney.

Bell, Diane and Klein, Renate (eds), (1996) *Radically Speaking: Feminism Reclaimed*, Spinifex, North Melbourne.

Bengelsdorf, C. and Hageman, A. (1979) 'Emerging from underdevelopment: women and work in Cuba' in *Capitalist Patriarchy and the Case for Socialist Feminism* (ed. Z. Eisenstein), 279–95. Monthly Review Press, New York.

Benston, Margaret (1969) 'The political economy of women's liberation', in *From Feminism to Liberation* (ed. E.H. Altbach),199–210. Shenkman, Cambridge, Mass.

Berger, John (1972) *Ways of Seeing*, BBC Corporation and Penguin, London.

Berman, Mildred (1977a) 'Facts and attitudes on discrimination as perceived by AAG members', *The Professional Geographer*, 29, 70–6.

Berman, Mildred (1977b) 'A response to the wolf who cried wolf', *The Professional Geographer*, 29 (4), 404–5.

Bhabha, Homi (1990) *Nation and Narration*, Routledge, London.

Bhabha, Homi (1994) *The Location of Culture*, Routledge, London & New York.

Binney, Judith (1987) 'Maori oral narratives, Pakeha written texts: two forms of telling history. *New Zealand Journal of History*, 21(1), 30–2.

Birch, Tony (1996) 'A land so inviting and still without inhabitants. Erasing Koori culture from (post-)colonial landscapes' in *Text, Theory, Space. Land, Literature and History in South Africa and Australia* (eds Darian-Smith, K.; Gunner, L. and Nuttall, S.), 173–85, Routledge, London & New York.

Bittman, Michael (1995) *Recent Changes in Unpaid Work*, Occasional Paper, ABS Catalogue No. 4154.0, Australian Government Publishing Service, Canberra.

Blainey, Geoffrey (1980) *A History of Camberwell*, Lothian Publishing Co., Melbourne and Sydney.

Bland, L.; Brunsdon, C; Hobson, D. and Winship, J. (1978) 'Women inside and outside the relations of production' in *Women Take Issue: Aspects of Women's Subordination* (eds Women's Studies Group), 35–77, Centre for Contemporary Cultural Studies, Hutchinson, London.

Blunt, Alison and Rose, Gillian (eds) (1994) *Writing, Women and Space: Colonial and Postcolonial Geographies*, The Guildford Press, New York.

Bondi, Liz (1990) 'Feminism, postmodernism, and geography: space for women?', *Antipode*, 22 (2), 156–67.

Bondi, Liz (1992) 'Gender and dichotomy', *Progress in Human Geography*, 16 (1), 98–104.

Bondi, Liz and Domosh, Mona (1992) 'Other figures in other places: on feminism, post-modernism and geography', *Environment and Planning D: Society and Space*, 10, 199–213.

Bordo, Susan (1993) *Unbearable Weight. Feminism, Western Culture and the Body*, University of California Press, Berkeley.

Bowlby, Sophie (1986) 'The place of gender in locality studies', *Area*, 18 (4), 327–31.

Broadbent, James (1987) 'The push east: Woolloomoolloo Hill, the first suburb, in *Sydney: City of Suburbs* (ed. M. Kelly), 12–29. NSW University Press, in Association with the Sydney History Group, Kensington, NSW.

Brownhill, Susan (1984) 'From critique to intervention', *Antipode*, 16 (3), 21–34.

Brownmiller, Susan (1976) *Against Our Will: Men, Women and Rape*, Simon and Schuster, New York.

Bruegel, Irene (1973) 'Cities, women and social class: A Comment', *Antipode*, 5, 62–3.

Bryson, Lois and Thompson, Faith (1972) *An Australian Newtown: Life and Leadership in a Working Class Suburb*, Penguin, Harmondsworth.

Bryson, Lois and Winter, Ian (1999) *Social Change, Suburban Lives. An Australian Newtown 1960s to 1990s*. Allen & Unwin, in association with the Institute of Family Studies, St Leonards.

Bryson, Valerie (1992) *Feminist Political Theory*, Paragon House, New York.

Burbidge, Andrew (1994) *Preliminary Calculations From the Australian Institute of Family Studies Family Living Standards Survey*, Melbourne.

Burbidge, Andrew and Winter, Ian (1996) 'Investigating urban poverty: the impact of state intervention and household change', *Urban Policy and Research*, 14 (2), 97–108.

Burgess, R. and Skeltys, N. (1992) 'The findings of the housing and location choice survey: an overview background paper', No. 11, *National Housing Strategy*, Australian Government Publishing Service, Canberra.

Burnett, Pat (1973) 'Social change, the status of women and models of city form and development', *Antipode*, 5, 57–61.

Butler, Judith (1990) *Gender Trouble: Feminism and the Subversion of Identity*, Routledge, New York.

Butler, Judith (1992) 'Contingent foundations: feminism and the question of "postmodernism"' in *Feminists Theorize the Political* (eds J. Butler and J. W. Scott), 3–21, Routledge, London & New York.

Butler, Judith (1993) *Bodies That Matter: On the Discursive Limits of 'Sex'*, Routledge, New York.

Cameron, Jenny (1996) 'Restructurings: Women's Employment, Households and Social Change' in *Restructuring Difference: Social Polarisation and the City* (K. Gibson et. al.), 13–28, Australian Housing Urban Research Institute, Working Paper No. 6, Melbourne.

Campioni, Mia and Gross, Elizabeth (1983) 'Love's labours lost: Marxism and feminism' in *Beyond Marxism? Interventions after Marx* (eds J. Allen and P. Patton), 113–141, Intervention Publications, Sydney.

Carby, H. (1982) 'White women listen! Black feminism and the boundaries of sisterhood' in *The Empire Strikes Back: Race and Racism in 70s Britain*, Centre for Contemporary Cultural Studies, Hutchinson, London.

Carter, Paul (1987) *The Road to Botany Bay*, Faber and Faber, London.

Castells, Manuel (1983) *The City and the Grassroots*, Arnold, London.

Castells, Manuel (1989) *The Informational City*, Blackwell, Oxford.

Charleville Times (1949) 'The Carnarvon Ranges fails as tourist attraction', 22 December.

Chesler, Phyllis (1972) *Women and Madness*, Doubleday, New York.

Chodorow, Nancy (1978) *The Reproduction of Mothering*, University of California Press, Berkeley.

Chow, Ray (1989) 'It's you and not me': domination and 'othering' in theorizing the 'Third World'" in *Coming to Terms: Feminism, Theory, Politics* (ed. E. Weed), Routledge, New York.

Chow, Ray (1996) 'Where have all the natives gone?' in *Postcolonial Theory A Reader*. (ed. P. Mongia), 122–47, London, Arnold.

Christopherson, Susan (1989) 'On being outside "the project"', *Antipode*, 21 (2), 83–9.

Clafton, Sara (1995) 'The $850 million landlords—Westfield Shopping Centres and the Lowry Family', *Business Review Weekly*, 8 May, 50–6.

Cliff, T. (1984) *Class Struggle and Women's Liberation*, Bookmarks, London.

Clifford, James (1988) *The Predicament of Culture*, Harvard University Press, Cambridge, Mass.

Clifford, James and Marcus, George E. (eds), (1986) *Writing Culture: the Poetics and Politics of Ethnography*, University of California Press, Berkeley.

Clough, Patricia Ticeneto (1995) *Feminist Thought: Desire, Power and Academic Discourse*, Blackwell, Oxford.

Cockburn, Cynthia (1981) 'The material of male power', *Feminist Review*, 9, 41–58.

Coher, Saul (1990) 'Geography–Gender Studies: fresh approaches but with integration', *Professional Geographer*, 42 (2), 228–31.

Connell Robert W. and Irving, Terry H. (1980) *Class Structure in Australian History*, Longman Cheshire, Melbourne.

Cooke, Philip (1984) 'Regions, class and gender: a European comparison', *Progress in Planning*, 22, 85–146.

Cooke, Philip (1988) 'Modernity, postmodernity and the city', *Theory, Culture and Society*, 5, 475–92.

Cooke, Philip (1989) *Localities: The Changing Face of Urban Britain*, Unwin Hyman, London.

Cooke, Philip (1990) *Back to the Future: Modernity, Postmodernity and Locality*, Unwin Hyman, London.

Cope, G.S. (1981) 'The Victorian woollen textile industry 1865–1881' Unpublished PhD thesis, University of Melbourne.

Corea, Gena (1985) *The Mother Machine: Reproductive Technologies from Artificial Insemination to Artificial Wombs*, Harper and Row, New York.

Coward, Rosalind (1983) *Patriarchal Precedents: Sexuality and Social Relations*, Routledge and Kegan Paul, London.

Coward, Rosalind (1984) *Female Desire: Women's Sexuality Today*, Paladin, London.

Crime Report (1998), 1 (5), June 2.

Crush, Jonathan (1994) 'Post-colonialism, de-colonization, and geography' in *Geography and Empire* (eds A. Godlewska and Smith, N.), 333–50. Blackwell, Oxford,

Curthoys, Ann (1994) 'Australian feminism since 1970' in *Australian Women. Contemporary Feminist Thought* (eds N. Greive and A. Burns), 14–28. Oxford University Press, Melbourne.

Daly, Mary (1978) *Gyn/ecology: the Meta-ethics of Radical Feminism*, Beacon Press, Boston.

Darian-Smith, Kate, Gunner, Liz and Nuttall, Sarah (eds), (1996) 'Introduction' in *Text, Theory, Space: Land, Literature and History in South Africa and Australia*, 1–20. Routledge, London and New York.

Davidoff, Leonore; L'Esperance, Jean and Newby, Howard (1976) 'Landscape with figures: home and community in English society' in *The Rights and Wrongs of Women* (eds J. Mitchell and A. Oakley), 139–75, Penguin, Harmondsmith.

Davis, Angela (1981) *Women, Race and Class*, Random House, New York.

Davis, Mike (1988) 'Urban Renaissance and the spirit of postmodernism' in *Postmodernism and Its Discontents* (ed. E.A. Kaplan), 79–87, Verso, London.

Davis, Mike (1990) *City of Quartz: Excavating the Future in Los Angeles*, Vintage, London.

Davison, Graeme (1978) *The Rise and Fall of Marvellous Melbourne*, Melbourne University Press, Melbourne.

Davison, Graeme (1994) 'The past and future of the Australian suburb' in *Suburban Dreaming: An Interdisciplinary Approach to Australian Cities*, (ed. L. C. Johnson), 99–113. Deakin University Press, Geelong.

Davison, Graeme and Dingle, Tony (1995) 'The view from the Ming wing' in *The Cream Brick Frontier: Histories of Australian Suburbia* (eds. G. Davison, T. Dingle and S. O'Hanlon), 2–17, Monash Publications in History No. 19, Clayton.

Davison, Graeme and Dingle, Tony and O'Hanlon, S. (eds), (1995) *The Cream Brick Frontier: Histories of Australian Suburbia*, Monash Publications in History No. 19, Clayton.

Dear, Michael (1986) 'Postmodernism and planning', *Environment and Planning D: Society and Space*, 4, 367–84.

Deatherage-Newsom, M. (1978) 'Teaching women's role in changing the face of the earth: how and why?', *Journal of Geography*, 82, 163–9.

Delphy, Christine (1976) *The Main Enemy*, Women's Research and Resources Centre, London.

Delphy, Christine (1984) *Close to Home: a Materialist Analysis of Women's Oppression*, Hutchinson in association with the Explorations in Feminism Collective, London.

Delmar, Rosalind (1979) 'Looking again at Engels's "Origin of the Family, Private Property and the State"' in *The Rights and Wrongs of Women* (eds J. Mitchell and A. Oakley), 271–87, Penguin, Harmonsdworth.

Denoon, Donald (1983) *Settler Capitalism: the Dynamics of Dependent Development in the Southern Hemisphere*, Clarendon Press, Oxford.

Department of Environment, Housing and Community Development (1978), *The Shopping Centre as a Community Leisure Resource*, Australian Government Publishing Service, Canberra.

Deutsche, Rosalyn (1991) 'Boy's town', *Environment and Planning D: Society and Space*, 9, 9–30.

Diamond, S. (1992) *Compromised Campus. The Collaboration of Universities with the Intelligence Community, 1945–1955*, Oxford University Press, New York.

Dinnerstein, Dorothy (1976) *The Mermaid and the Minotaur: Sexual Arrangements and Human Malaise*, Harper and Row, New York.

Dirlik, Arif (1996) 'The postcolonial aura: Third World criticism in the age of global capitalism' in *Postcolonial Theory. A Reader.* (ed. P. Mongia), 294–320, Arnold, London.

Domosh, Mona (1991) 'Towards a feminist historiography of geography', *Transactions*, Institute of British Geographers NS 16, 95–104.

Dovey, Kim (1994) 'Dreams on display: suburban ideology in the model home' in *Beasts of Suburbia* (eds S. Ferber, C. Healy and C. McAuliffe), 127–47, Melbourne University Press, Melbourne.

Dowling, Robyn (1996) 'Making a place for the working class? Early visions of Mt Druitt, NSW' in *Gender and Environments Proceedings of the Institute of Australian Geographer*, Gender and Geography Study Group Conference (ed. Conference Organising Committee and I. Hay), 34–42. Adelaide.

Driver, Felix (1992) 'Geography's empire: histories of geographical knowledge', *Environment and Planning* 10, 23–40.

Duncan, James (1993) 'Sites of representation: place, time and the discourse of the other' in *Place/Culture/Representation* (eds J. Duncan and D. Ley), 39–56. Routledge, London.

Duncan, N. and Sharp, J.P. (1993) 'Confronting representation(s)', *Environment and Planning D*, 11. 473–86.

Dworkin, Andrea (1976) *Our Blood: Prophesies and Discourse on Sexual Politics*, Harper and Row, New York.

Dworkin, Andrea (1981) *Pornography: Men Possessing Women*, G.P. Putnams, New York.

Dyck, Isabel (1989) 'Integrating home and workplace: women's daily lives in a Canadian suburb', *Canadian Geographer* 33 (4), 329–41.

Eade (1997) *Living the Global City: Globalization as a Local Process*, Routledge, London and New York.

Einhorn, B. (1981) 'Socialist emancipation: the women's movement in the German Democratic Republic', *Women's Studies International Forum*, 4 (4), 435–52.

Eisen, A. (1984) *Women and Revolution in Vietnam*, Zed Press. London.

Eisenstein, Hester (1984) *Contemporary Feminist Thought*, Unwin Paperbacks, London and Sydney.

Eisenstein, Zillah R. (1979) 'Developing a theory of capitalist patriarchy and socialist feminism' in *Capitalist Patriarchy and the Case for Socialist Feminism*, (ed. Z. Eisenstein), 5–40. Monthly Review Press, New York.

Eisenstein, Zillah R. (1981) *The Radical Future of Liberal Feminism*. Unwin Paperbacks, London.

Elder, Glen (1995) 'Of moffies, kaffirs and perverts: male homosexuality and the discourse of moral order in the apartheid state' in *Mapping Desire* (eds David Bell and Gill Valentine), 56–65, Routledge, London.

Elkin, Adolphus (1939) 'Introduction' in *Aboriginal Women: Sacred and Profane*. (ed. P. Kaberry), Routledge, London.

Engels, Friedrich (1975) *The Origin of the Family, Private Property and the State,* International Publishers, New York.

England, Kim (1994) 'Getting personal: reflexivity, positionality and feminist research', *The Professional Geographer* 46 (1), 80–9.

Ericksen, Julia (1977) 'An analysis of the journey to work for women', *Social Problems* 24, 428–35.

Fagan, Robert. (1986) 'Industrial restructuring and the metropolitan fringe: Growth and disadvantage in Western Sydney', *Australian Planner*, 24 (1), 11–17.

Fagan, Robert H. and Webber, Michael. (1999) *Global Restructuring. The Australian Experience*, Oxford University Press, Sydney.

Fahey, Stephanie (1988) 'Putting gender into geography', *Australian Geographical Studies*, 26 (1), 202–13.

Fainstein, S., Gordon, I. and Harloe, M. (1993) *Divided Cities*, Basil Blackwell, Oxford.

Fanon, Franz (1968) *The Wretched of the Earth*, Grove, New York.

Featherstone, Michael (1991) *Consumer Culture and Postmodernism*, Sage Publications, London.

Fincher, Ruth (1993) 'Women, the state and the life course in urban Australia' in *Full Circles: Geograpies of Women Over the Life Course* (eds C. Kate and J. Monk), 243–63, Routledge, London.

Fincher, Ruth and Jacobs, Jane (eds) (1998) *Cities of Difference*, The Guildford Press, London.

Firestone, Shulamith (1979) *The Dialectic of Sex*, The Women's Press, London.

Fiske, John, Hodge, Bob and Turner, Graham (1987) *Myths of Oz. Reading Australian Popular Culture*, Allen & Unwin, North Sydney.

Foord, Jo (1980) 'Women's place—women's space: a comment', *Area*, 12 (1), 49–50.

Foord, Jo and Gregson, Nicky (1986) 'Patriarchy: towards a reconceptualisation', *Antipode* 18 (2), 186–211.

Forster, Clive (1986) 'Economic restructuring, urban policy and patterns of deprivation in Adelaide', *Australian Planner*, Vol. 24 (1), 6–10.

Forster, Clive (1995) *Australian Cities: Continuity and Change*, Oxford University Press, Melbourne.

Foucault, Michel (1972) *The Archaeology of Knowledge and the Discourse of Language*, Patheon, New York.

Foucault, Michel (1973) *The Birth of the Clinic: an Archaeology of Medical Perception*, Pantheon Books, New York.

Foucault, Michel (1977) *Discipline and Punish. The Birth of the Prison*, Penguin, Harmondsmith.

Foucault, Michel (1979) *The History of Sexuality, Vol. 1, An Introduction*, Penguin, Harmondworth.

Fox, Bonnie (1986). 'Never done: The struggle to understand domestic labour and women's oppression' in *The Politics of Diversity* (eds R. Hamilton and M.Barrett), 180–9, Verso, London.

Fraser, Nancy (1997) 'Equality, difference and democracy: recent feminist debates in the United States' in *Feminism and the New Democracy* (ed. J. Dean), 98–109, Sage, London.

Freestone, Robert (1989) *Model Communities: the Garden City Movement in Australia*, Nelson, Melbourne.

Friedan, Betty (1963) *The Feminine Mystique*, Harmondsworth, Penguin.

Friedman, John (1986), 'The world city hypothesis', *Development and Change* 17, 69–74.

Frost, Lionel (1994) 'Nineteenth century Australian cities', in *Suburban Dreaming. An Interdisciplinary Approach to Australian Cities* (ed. L.C. Johnson), 22–32. Deakin University Press, Geelong.

Frost, Lionel and Dingle, Tony (1995) 'Sustaining suburbia: An historical perspective on Australia's urban growth' in *Australian Cities* (ed. P. Troy), 20–38. Cambridge University Press, Cambridge.

Game, Ann Pringle, Rosemary (1979) 'Sexuality and the suburban dream', *Australian and New Zealand Journal of Sociology*, 15 (2), 4–15.

Game, Ann and Pringle, Rosemary (1983) *Gender at Work*, George Allen & Unwin, Sydney.

Garden, Don (1995) 'Type 15, Glengarry and Catalina: The changing space of the A.V. Jennings home in the 1960s' in *The Cream Brick Frontier: Histories of Australian Suburbia* (eds G. Davison, T. Dingle and S. O'Hanlon), 140–53, Monash Publications in History No. 19, Clayton, Vic.

Gardner, C. and Sheppard, J. (1989) *Consuming Passion: the Rise of Retail Culture*, Unwin Hyman, London.

Gatens, Moira (1992) *Feminism and Philosophy: Perspectives on Difference and Equality*, Polity Press, Cambridge.

Geelong Advertiser (1991) '850 locals will lose Ford jobs', January 25.

Geelong Regional Commission (1981) *Employment Strategy for the Geelong Region*, GRC, Geelong.

Gibbs, Anna (1985) 'Contemporary feminism' in *Women and Social Change HUX 152* (Women's Studies Course Team), 83–102. Deakin University, Geelong.

Gibson, Katherine and Graham, Julie (1992) 'Rethinking class in industrial geography: Creating a space for an alternative politics of class', *Economic Geography*, 68 (2), 109–27.

Gibson, Katherine (1996) 'Social polarisation and the politics of difference: Discourses in collision or collusion' in *Restructuring Difference: Social Polarisation and the City* (K. Gibson et. al.), 5–12, Australian Housing Urban Research Institute, Working Paper No. 6, Melbourne.

Gibson-Graham, J.K. (1996) *The End of Capitalism (As We Knew It)*, Blackwell, Oxford.

Gibson, Katherine et. al (1996) *Restructuring Difference: Social Polarisation and the City*, Australian Housing and Urban Research Institute Working Paper No. 6, AHURI, Melbourne.

Gier, Jaclyn and Walton, John (1987) 'Some problems with reconceptualising patriarchy', *Antipode*, 19 (1), 54–8.

Gilbert, Kevin (1977) *Living Black*, Alan Lane/Penguin, Melbourne.

Golledge, Reginald and Halperin, W. C. (1983) 'On the status of women in geography', *The Professional Geographer*, 35, 214–18.

Gordon, M. and Riger, S. (1989) *The Female Fear*, The Free Press, New York.

Goss, Jon (1992) 'Modernity and post-modernity in the retail landscape' in *Inventing Places. Studies in Cultural Geography* (eds K. Anderson and F. Gale), 159–77, Longman Cheshire, Melbourne.

Grace, Helen; Hage, Ghassan; Johnson, Lesley; Langsworth, Julie and Symonds, Michael (1997) *Home/Worlds: Space, Community and Marginality in Sydney's West*, Pluto Press, Sydney.

Graham, Julie (1992) 'Post-Fordism as politics: The political consequences of narratives on the left', *Environment and Planning D: Society and Space*, 10, 393–410.

Gregory, Derek (1994) *Geographical Imaginations*, Basil Blackwell, Cambridge, Mass.

Gregson, Nicky and Foord, Jo (1987) 'Comments on Critics', *Antipode*, 19 (3), 371–5

Gregson, Nicky and Lowe, M. (1994) *Servicing the Middle Class: Gender and Waged Domestic Labour in Contemporary Britain*, Routledge, London and New York.

Griffin, Susan (1978) *Women and Nature: The Roaring Inside Her*, Harper and Row, New York.

Griffin, Susan (1979) *Rape: The Power of Consciousness*, Harper and Row, San Francisco.

Griffin, Susan (1981) *Pornography and Silence: Culture's Revenge Against Nature*, Harper and Row, New York.

Grosz, Elizabeth (1988) 'Sexual difference and the problem of essentialism', *Inscriptions* 5, 86–101.

Grosz, Elizabeth (1993) *Sexy Bodies: The Strange Canalities of Feminism*, Routledge, London and New York.

Grosz, Elizabeth (1994) *Volatile Bodies: Towards a Corporeal Feminism*, Allen & Unwin, St Leonards, NSW.

Gruen, Victor (1973) *Centres for the Urban Environment: Survival of the Cities*, Van Nostrand Reinhold Co., New York.

Guha, Ranajit (1982–87) *SubAltern Studies: Writings on South Asian History and Society*, 5 volumes, Oxford University Press, Oxford.

Gunew, Sneja (1983) 'Migrant women writers: who's on whose margins?' *Meanjin* 42, 16–26.

Gunew, Sneja (1990) 'Denaturalizing cultural nationalisms: Multicultural readings of "Australia"' in *Nation and Narration*, 99–120, (ed. H. Bhabha) Routledge, London.

Gunew, Sneja (1991) 'Margins: acting like a (foreign) woman', *Hecate*, 17, 31–5.

Gunew, Sneja (1993) 'Feminism and the politics of irreducible difference: multiculturalism/ethnicity/race' in *Feminism and the Politics of Difference* (eds S. Gunew & A. Yeatman), 1–19, Allen & Unwin, St Leonards, NSW.

Gunew, Sneja and O'Longley, Katarina (eds) (1988) *Striking Chords: Multicultural Literary Interpretations*

Gunew, Sneja and Mahyuddin, Jan (1992) *A Bibliography of Australian Multicultural Writers*, Centre for Studies in Literary Education, Geelong, Vic.

Gunew, Sneja and Yeatman, Anna (eds) (1993) *Feminism and the Politics of Difference*, Allen & Unwin, Sydney.

Hall, Peter (1984/1966) *The World Cities*, Wedernfeld and Nicholson, London.

Hall, Stuart (1987) 'Minimal selves', *Identity Documents*, 6, 44–6.

Hall, Stuart (1990) 'Cultural identity and diaspora' in *Identity: Community, Culture, Difference*, 222–217, (ed. J. Rutherford), Kawrence and Wishart, London.

Hall, Stuart (1991) 'The local and the global: globalization and identity' in *Culture, Globalization and the World-System: Contemporary Conditions for the Representation of Identity* (ed. A.D. King), Macmillan, Basingstoke.

Hall, Stuart (1992) 'New ethnicities' in *Race', Culture and Difference* (eds J. Donald and A. Rattansii), Sage, London.

Hall, Stuart (1993) 'Culture, community, nation', *Cultural Studies* 7 (3), 349–63.

Hamilton, A. (1981) 'A complex strategical situation: gender and power in Aboriginal Australia' in *Australian Women: Feminist Perspectives* (eds N. Grieve & P. Grimshaw), 69–85, Oxford University Press, Melbourne.

Hamlyn-Harris, M. (n.d.) 'Carnarvon Journey and other Verses of the Series', *Mid Highways and Byways of Queensland*, Tribute to D.A. O'Brien, FRGSA. RGSA (Qld) Records.

Hanmer, J. and Maynard, Mary. (1984) *Well Founded Fear: A Community Study of Violence to Women*, Hutchinson, London.

Hanson, Susan and Pratt, Geraldine (1995) *Gender, Work and Space*, Routledge, London.

Harley, J.B. (1989) 'Deconstructing the map', *Cartographica* 26, 1–20.

Hartmann, Heidi (1981) 'The unhappy marriage of Marxism and feminism: Towards a more progressive union' in *Women and Revolution* (ed. L. Sargent), 1–41, South End Press, Boston.

Harvey, David (1983) 'Owen Lattimore: A Memoir' *Antipode* 15, 3–11.

Harvey, David (1989a) *The Condition of Postmodernity*, Basil Blackwell, Oxford.

Harvey, David (1989b) 'Flexible accumulation through urbanization: reflections on "postmodernism" in the American city' in *The Urban Experience* (D. Harvey), Basil Blackwell, Oxford.

Hawkins, Gay and Gibson, Katherine (1994) 'Cultural planning in Australia: Policy dreams, economic realities' in *Metropolis Now* (eds K. Gibson and S. Watson), 217–28, Pluto, Sydney.

Hay, Iain (1995) 'The strange case of Dr Jeckyll on Hyde Park: Fear, media and the conduct of emancipatory geography', *Australian Geographical Studies* 33 (2), 257–71.

Hayden, Dolores (1981) *The Grand Domestic Revolution: A History of Feminist Designs for American Houses*, MIT Press, Massachusetts and London.

Hayford, Alison (1974) 'The geography of women: An historical introduction', *Antipode*, 6, 1–13.

Hebdidge, D. (1989) 'New times: After the masses' *Marxism Today*, January, 48–53.

Hodge, Stephen (1995) '"No fags out there" Gay men, identity and suburbia', *Journal of Interdisciplinary Gender Studies*, 1 (1), 41–8.

Hodge, Stephen (1996) 'Fluid or fixed? Theorising and performing the relationship between sexuality and space' in *Gender and Environments*. Proceedings of the Institute of Australian Geographers' Gender and Geography Study Group Conference, (eds Conference Organising Committee and Iain Hay), 78–86, Adelaide.

hooks, bell (1982) *Ain't I a Woman? Black Women and Feminism*, Pluto, London.

hooks, bell (1989) *Talking Back: Thinking Feminist, Thinking Black*, South End Press, Boston.

hooks, bell (1990) *Yearning: Race, Gender and Cultural* Politics. Between the Lines, Toronto.

Horacek, Judy (1997) *Woman with Altitude*, Hodder Headline, Rydalmere, NSW.

Hoskins, Ian (1994) 'Constructing time and space in the garden suburb' in *Beasts of Suburbia* (eds S. Ferber, C. Healy and C. McAuliffe), 1–17, Melbourne University Press, Melbourne.

Hourani, Pam (1990) 'Spatial organisation and the status of women in nineteenth century Australia', *Australian Historical Archaeology*, 8, 70–7.

Howe, Anna and O'Connor Kevin (1982) 'Travel to work and labour force participation of men and women in an Australian metropolitan area', *Professional Geographer*, 34, 50–64.

Howe, Anna (1984) 'Conference report from a professional woman geographer with an interest in education', *Australian Geographical Studies*, 22 (1), 151.

Howe, Renate (1994) 'Inner suburbs: from slums to gentrification, in *Suburban Dreaming* (ed. L.C. Johnson), 141–59, Deakin University Press, Geelong.

Howe, Renate (1995) 'The concrete house frontier: The Victorian Housing Commission and the planning of Melbourne in the 1940s and 1950s' in *The Cream Brick Frontier: Histories of Australian Suburbia* (eds G. Davison, T. Dingle and S. O'Hanlon), 74–87, Monash Publications in History No. 19, Clayton.

Hudson, B. (1977) 'The new geography and the new imperialism, 1870–1918', *Antipode 9*, 12–19.

Huggins, Jackie (1987a) 'Black women and women's liberation', *Hecate*, 13, 5–23.

Huggins, Jackie (1987b) '"Firing on in the mind": Aboriginal women domestic servants in the inter war years', *Hecate* 13, 5–23.

Huggins, Jackie (1992) 'A contemporary Aboriginal woman's relationship to the white women's movement' in *A Woman's Place in Australia* HUA 713 (A Woman's Place Course Team), 16–26. Deakin University, Geelong.

Huggins, Jackie and Blake, Thom (1992) 'Protection or persecution? Gender relations in the era of racial segregation' in *Gender Relations. Domination and Negotiation* (eds K. Saunders and R. Evans), 4–58. Harcourt Brace and Jovanovich, Sydney.

Huxley, Margo (1994) 'Space, knowledge, power and gender' in *Suburban Dreaming: An Interdisciplinary Approach to Australian Cities* (ed. L.C. Johnson), 181–92. Deakin University Press, Geelong.

Ingram, Gordon Brent; Bouthillette, Anne-Marie and Retter, Yolanda (eds) (1997) *Queers in Space*, Bay Press, Seattle, Wash.

Jackson, Peter (1991) 'The crisis of representation and the politics of position', *Environment and Planning D: Society and Space*, 9, 131–4.

Jacobs, Jane (1996) *Edge of Empire*, Routledge, London & New York.

Jameson, Frederic (1984) 'The cultural logic of late capitalism', *New Left Review*, vol. 146, 53–92.

Jeffreys, Elaine (1991) 'What is "difference" in feminist theory and practice?', *Australian Feminist Studies*, 14, Summer, 1–13.

Jencks, Charles (1986) *What is post-modernism?*, Paper given to Post-Modernism Conference, North Western University, Evanston, Illinois.

Johnson, Louise (1984) *Gaslight Sydney*, George Allen & Unwin, Sydney.

Johnson, Louise (1985) 'Gender, gene(r)ics and the possibility of feminist geography', *Australian Geographical Studies*, 23(1), 161–71

Johnson, Louise (1987) '(Un) Realist perspectives: patriarchy and feminist challenges in geography', *Antipode*, 19 (2) 210–15.

Johnson, Louise (1989) *Embodying Geography—Some Implications of Considering the Sexed Body in Space*. Proceedings of the 24th New Zealand Geographical Society Conference University of Otago, Dunedin, pp. 134–8.

Johnson, Louise C. (1990a) 'Socialist feminisms' in *Feminist Knowledge: Critique and Construct* (ed. S. Gunew), 304–31, Routledge, London.

Johnson, Louise C. (1990b) 'New patriarchal economies in the Australian textile industry', *Antipode*, 22 (1), 1–32.

Johnson, Louise C. (1991) 'The Australian textile industry, 1865–1990. A feminist geography', Unpublished PhD thesis, Monash University, Melbourne.

Johnson, Louise (1992) 'Housing desire: A feminist geography of suburban housing', *Refractory Girl*, 42, 40–6.

Johnson, Louise (1994a) 'A (post) modern suburb?: Lynch's Bridge in Melbourne' in *Beasts of Suburbia* (eds S. Ferber, C. Healy and C. McAuliffe), 170–84. Melbourne University Press, Melbourne.

Johnson, Louise C. (1994b) 'The postmodern Australian city' in *Suburban Dreaming: An Interdisciplinary Approach to Australian Cities* (ed. L.C. Johnson), 51–72. Deakin University Press, Geelong.

Johnson, Louise (1994c) 'Occupying the suburban frontier: accommodating difference on Melbourne's urban fringe' in *Writing Women and Space: Colonial and Postcolonial Geographies* (eds A. Blunt and G. Rose), 141–68, Guildford Press, New York.

Johnson, Louise (1994d) 'Colonising the suburban frontier: place-making on Melbourne's urban fringe' in *Metropolis Now*, (eds K. Gibson and S. Watson), 46–59, Pluto Press, Sydney.

Johnson, Louise (1996a) 'Restructuring and socio-economic polarisation in a regional industrial centre' in *Restructuring Difference: Social Polarisation and the City* (K. Gibson et al.), 43–57. Australian Housing Urban Research Institute, Working Paper No. 6, Melbourne.

Johnson, Louise (1996b) 'Refashioning Melbourne's west' in *Gender and Environments*, Proceedings of the Institute of Australian Geographers' Gender and Geography Study Group Conference, (eds Conference Organising Committee and Iain Hay), 57–77, Adelaide.

Johnson, Louise C. (1997) 'The oracles of Delfin: women and suburban development', *Urban Policy and Research*, 15 (2), June: 25–36.

Johnson, Louise C. (1999) 'Powerlines: A Cultural Geography of Domestic Open Space' in *Australian Cultural Geographies*, (ed E. Stratford), 87–108, Oxford University Press, Melbourne.

Johnston, Linda (1995) 'The politics of the pump. Hard core gyms and women body builders', *New Zealand Geographer*, 51 (1), 16–18.

Johnson, Pauline (1991) 'Feminism and liberalism', *Australian Feminist Studies* 14, Summer, 57–68.

Johnston, R.J. and Brack, E.V. (1983) 'Appointment and promotion in the academic labour market: A preliminary survey of British university departments of geography', *Transactions of the Institute of British Geographers*, NS, 8, 100–11.

Jury, A (1983) 'The drought eases for International Harvester', *Australian Business*, June 2, 94–100.

Kass, Terry (1987) 'Cheaper than rent: Aspects of the growth of owner-occupation in Sydney 1911–1966' in *Sydney: City of Suburbs* (ed. M. Kelly), 77–94, NSW University Press in association with the Sydney History Group, Sydney.

Katz, Cindi (1994) 'Playing the field: questions of fieldwork in geography', *The Professional Geographer*, 46 (1), 67–72.

Kazi, H. (1986). 'The beginning of a debate long overdue: some observations on ethnocentrism and socialist feminist theory', *Feminist Review* 22, 87–91.

Keith, Michael and Pile, Steve (eds) (1993) *Place and the Politics of Identity* Routledge, London & New York.

Kemeny, Jim (1981) *The Myth of Home Ownership*, Routledge and Kegan Paul, London.

Kemeny, Jim (1988) 'The ideology of home ownership' in *Urban Planning in Australia: Critical Readings* (eds J. Brian McLoughlin and M. Huxley), 251–8, Longman Cheshire, Melbourne.

King, Anthony (1990) *Global Cities: Post-imperialism and the Internationalisation of London*, Routledge, London.

Kingston, Beverley (1994) *Basket, Bag and Trolley: A History of Shopping in Australia*, Oxford University Press, Melbourne.

Kirby, Stewart (1996) 'Gay men's perceptions and experiences of everyday space and place' in *Gender and Environments*, Proceedings of the Institute of Australian Geographers' Gender and Geography Study Group Conference (eds Conference Organising Committee and Iain Hay), 87–104, Adelaide.

Klein, Renate (1989) *Infertility: Women Speak Out About Their Experiences of Reproductive Medicine*, Pandora, London.

Knopp, Lawrence and Lauria, Micky (1987) 'Gender relations as a particular form of social relations', *Antipode*, 19 (1), 48–53.

Knopp, Lawrence (1990) 'Some theoretical implications of gay involvement in the urban land market', *Political Geography Quarterly*, 9 (4), 337–52.

Knopp, Lawrence (1992) 'Sexuality and the spatial dynamics of capitalism', *Environment and Planning D: Society and Space*, 10, 651–69.

Lack, John (1991) *A History of Footscray*, Hargreen, Melbourne.

Lancaster Regionalism Group (1985) *Localities, class and gender*, Pion, London.

Langford, Ruby Ginibi (1994) *My Bundalung People*, University of Queensland Press, Brisbane.

Langford, Ruby (1988) *Don't Take Your Love to Town*, Penguin, Ringwood, Vic.

Langton, Marcia (1988) 'The getting of power', *Australian Feminist Studies* 6, 1–5.

Langton, Marcia (1990) *Feminism: What do Aboriginals Gain?*, Broadside Newsletter of the National Foundation of Australian Women. 1 (3).

Lauria, Micky and Knopp, Laurence (1985) 'Towards an analysis of the role of gay communities in the urban renaissance', *Urban Geography*, 6, 152–69.

Lash, Scott and Urry, John (1987) *The End of Organized Capitalism*, Polity Press, Cambridge.

de Lauretis, Teresa (1988) 'Displacing hegemonic discourses: reflections on feminist theory in the 1980s', *Inscriptions* 3/4, 127–44.

Lawrence, Roderick (1979) 'The organization of domestic space', *Ekistics*, 46 (275), 135–40.

Leach, B., Lesiuk, E. and Morton, P.E. (1986) 'Perceptions of fear in the urban environment', *Women and Environments*, Spring, 10–12.

Lee, David (1978) 'Feminist approaches to teaching geography', *Journal of Geography*, 77 (5), 180–3.

Lee, David (1990) 'The status of women in geography: things change, things remain the same', *The Professional Geographer*, 42, 202–11.

Leslie, D. A. (1993) 'Femininity, post-Fordism and the "new traditionalism" ', *Environment and Planning D: Society and Space*, 11 (6), 689–708.

Letter, RGSA (Q) (1956) Circular letter from Peter Grant and Tom Denning, Joint Organisers of the 1956 RGSA (Qld) Carnarvon Expedition, to prospective participants. RGSA (Qld) Records.

Leunig, Mary (1982) *There's No Place Like Home: Drawings by Mary Leunig*, Penguin, Melbourne

Lewis, Jane (1984) 'The role of female employment in the industrial restructuring and regional development of the United Kingdom', *Antipode*, 16 (3), 47–59.

Linge, Godfrey (1979) *Industrial Awakening: A Geography of Australian Manufacturing, 1788–1890*, ANU Press, Canberra.

Livingstone, David N. (1992) *The Geographical Tradition*, Blackwells, Oxford.

Lloyd, Bonnie and Rengert, A. (1977) 'Women in geography curriculum: an introduction to the issue', *The Journal of Geography*, 77 (5), 164–5.

Lloyd, Genevieve. (1984) *The Man of Reason: 'Male' and 'Female' in Western Philosophy*, Methuen, London.

Longhurst, Robyn (1994) 'The geography closest in—the body. The politics of pregnability', *Australian Geographical Studies*, 32 (2), 214–23.

Longhurst, Robyn (1995) 'Discursive constraints on pregnant women's particpation in sport', *New Zealand Geographer*, 51 (1), 13–15.

Lopata, Helena Z. (1980) 'The Chicago woman: A study of patterns of mobility and transport', *Signs* 5 (3), Supplement, S161–169.

Lunt, P. and Livingstone, S. M. (1992) *Mass Consumption and Personal Identity*, Oxford University Press, Buckinghamshire.

Lynch, G. and Atkins, S. (1988) 'The influence of personal security fears on women's travel patterns', *Transportation*, 15, 275–77.

Lyotard, Jean Francois (1984) *The Postmodern Condition: A Report on Knowledge*, University of Minnesota Press, Minneapolis.

MacIntyre, Clement (1991) '..."now you're in the family zone": Housing and domestic design in Australia', *Journal of Australian Studies*, 30, 58–71.

Mackenzie, Suzanne (1980) 'Women's place—women's space', *Area*, 12 (1), 47–9.

Mackenzie, Suzanne (1987) 'Neglected spaces in peripheral places: Homeworkers and the creation of a new economic centre', *Cahiers de Geographie du Quebec*, 31 (83), 247–60.

Mackenzie, Suzanne and Rose, Damaris (1983) 'Industrial change, domestic economy and home life' in *Redundant Spaces? Studies in Industrial Decline and Social Change* (eds J. Anderson, S. Duncan and R. Hudson), 155–200, Academic Press, London.

Mani, L. (1990) 'Multiple mediations: feminist scholarship in the age of multinational reception', *Feminist Review*, 35, 4–41.

Marcus, George E. (ed.) (1992) *Rereading Cultural Anthropology*, Duke University Press, Durham and London.

Marcuse, Peter (1995) *Is Australia Different? Globalization and the New Economy*, Australian Housing and Urban Research Institute (AHURI), Working Paper No. 4. AHURI, Melbourne.

Marsh, B. (1985) *A Corporate Tragedy: the Agony of International Harvester*, Doubleday, New York.

Marx, Karl and Engels, Friedrich (1976) *Karl Marx, Friedrich Engels: Collected Works.Vol. 5*, Lawrence and Wishart, London.

Massey, Doreen (1984). *Spatial Divisions of Labour: Social Structures and the Geography of Production*, London, Macmillan.

Massey, Doreen (1991) 'Flexible sexism', *Environment and Planning D: Society and Space*, 9, 31–57.

Massey, Doreen (1994) *Space, Place and Gender*, Polity, Cambridge.

Matthews, Jill J. (1984) *Good and Mad Women: The Historical Construction of Femininity in Twentieth Century Australia*, George Allen & Unwin, Sydney.

Mathews, John (1989) *Tools of Change. New Technologies and the Democratisation of Work*, Pluto, Leichhardt, NSW.

Matrix (1984) *Making Space: Women and the Man Made Environment*, Pluto, London.

Mawson Lakes Update (Nov–Dec 1998) Issue 4.

Mawson Lakes (1998/99) *Summer 98/99 Catalogue*, Delfin Lend Lease, Adelaide.

Mazey, M. and Lee, David (1983) *Her Space. Her Place*. Association of American Geographers, Resource Publications in Geography, Washington, DC.

McCarty, John (1978) 'Australian capital cities in the nineteenth century' in *Australian Capital Cities* (eds J.W. McCarty and C.B. Schedvin), 9–25. Sydney University Press, Sydney.

McDonough, R. and Harrison, R. (1978) 'Patriarchy and relations of production' in *Feminism and Materialism: Women and Modes of Production* (eds A. Kuhn and A. M. Wolpe), 11–41. Routledge and Kegan Paul, London.

McDowell, Linda (1979) 'Women in British geography', *Area*, 11, 151–4.

McDowell, Linda (1983) 'Towards an understanding of the gender division of urban space', *Environment and Planning D: Society and Space* 1, 59–72.

McDowell, Linda (1986) 'Beyond patriarchy: A class-based explanation of women's subordination', *Antipode*, 18 (3), 311–21.

McDowell, Linda (1990) 'Sex and power in academia', *Area*, 22 (4), 323–32.

McDowell, Linda (1991a) 'Life without Father and Ford: The New Gender Order of Post-Fordism', *Transactions of the Institute of British Geographers*, N.S, 16, 400–19.

McDowell, Linda (1991b) 'The baby and the bath water: diversity, deconstruction and feminist theory in geography', *Geoforum*, 22 (2), 123–33.

McDowell, Linda and Massey, Doreen (1984) 'A woman's place?' in *Geography Matters! A Reader* (eds D. Massey and J. Allen), 128–47, Cambridge University Press, Cambridge.

McIntosh, Mary (1982) 'The family in socialist-feminist politics' in *Feminism, Culture and Politics* (eds R. Brunt and C. Rowan), 109–29, Lawrence and Wishart, London.

Mee, Kathleen (1994) 'Dressing up the suburbs: Representation of western Sydney' in *Metropolis Now* (eds K. Gibson and S. Watson), 60–77, Pluto Press, Sydney.

Melbourne Star (1935) 'Geelong is now world famous as the "Southern Bradford"', 2 December.

Milkman, Ruth (1982) 'Redefining "women's work": The sexual division of labour in the auto industry during World War II', *Feminist Studies*, 8 (2), 336–72.

Mill, John Stuart and Taylor, Harriet (1869/1983) *The Subjection of Women*, Virago, London.

Miller, Daniel, Jackson, Peter, Thrift, Nigel, Holbrook, Beverley and Rowlands, Michael (1998) *Shopping, Place and Identity*, Routledge, London.

Millett, Kate (1977) *Sexual Politics*, Virago, London.

Minh-ha, Trinh (1989) *Woman, Native, Other*, Indiana University Press, Bloomington, Ind.

Mirza, H.S. (1986) 'The dilemma of socialist-feminism: a case for black feminism', *Feminist Review* 22, 103–5.

Mitchell, Juliet (1966) 'Women: The longest revolution', *New Left Review*, 40, 11–37.

Mitchell, Juliet (1971) *Woman's Estate*, Vintage Books, New York.

Mitchell, Juliet (1979) *Psychoanalysis and Feminism*, Penguin, Harmondsworth.

Mohanty, Chandra (1988) 'Under western eyes: feminist scholarship and colonial discourse', *Feminist Review* 30, 61–88.

Moi, Toril (1985) *Sexual/Textual Politics*, Methuen, London.

Momsen, Janet H. (1980) 'Women in Canadian geography', *Professional Geographer*, 32 (3), 365–9.

Momsen, Janet H. and Townsend, J. (eds) (1987) *Geography of Gender in the Third World*, Hutchinson, London.

Mongia, P. (1996) *Postcolonial Theory. A Reader*, Arnold, London.

Monk, Janice and Hanson, Susan (1982) 'On not excluding half of the human in human geography', *Professional Geographer*, 34 (1), 11–23.

Monk, Janice (1984) 'Integrating women in the geography curriculum', *Journal of Geography*, 82, 271–3.

Monk, Janice (1992) 'Gender in the landscape: expressions of power and meaning' in *Inventing Places: Studies in Cultural Geography*, (eds K. Anderson and F. Gale), 123–38, Longman Cheshire, Melbourne.

Monk, Janice (1994) 'Place matters: comparative international perspectives on feminist geography', *The Professional Geographer* 46 (3), 277–88.

Moraga, C. and Anzaldua, G. (1981) (eds) *This Bridge Called My Back: Writings by Radical Women of Color*, Persephone Press, Watertown, Mass.

Morgan, Sally (1987) *My Place*, Fremantle Arts Centre Press, Fremantle, WA.

Morris, Meaghan (1988) 'Things to do with shopping centres', in *Grafts. Feminist Cultural Criticism*, (ed. S. Sheridan), 193–227, Verso, London.

Murgatroyd, Linda (1985) 'Occupational stratification and gender' in *Localities, Class and Gender*, (The Lancaster Regionalism Group), 121–43, Pion, London.

Murphy, Peter and Watson, Sophie (1997) *Surface City: Sydney at the Millenium*, Pluto, Sydney.

Narogin, Mudrooroo (1990) *Writing from the Fringe*, Hyland House, Melbourne.

Nash, Kate (1998) *Universal Difference: Feminism and the Liberal Undecidability of Women*, Macmillan, Houndmills.

Newman, R.P. (1992) *Owen Lattimore and the 'Loss' of China*, University of California Press, Berkeley.

Nicholson, Linda. (ed.) (1990) *Feminism/Postmodernism*, Routledge, New York and London.

O'Brien n.d., cited in Anon., p. 1, (n.d.) 1. Unpublished manuscript, no title. Records of the RGSA (Qld).

O'Brien, Danny (1942) *Carnarvons*, published under the auspices of the Carnarvon Range Association, Brisbane. Records of the RGSA(Qld).

O'Brien, Danny (1950) *The Carnarvon Ranges: Expedition of 31 all told, 18.9.'50 to 3.11.'50*, Unpublished manuscript. Records of the RGSA (Qld). 10

O'Brien, Mary (1981) *The Politics of Reproduction*, Routledge and Kegan Paul, London.

O'Brien, Mary (1982) 'Feminist theory and dialectical logic', *Signs*, 7 (1), 144–57.

Okin, Susan M. (1979) *Women in Western Political Thought*, Princeton University Press, Princeton.

Ong, A. (1988) 'Colonialism and modernity: feminist representations of women in non-Western societies', *Inscriptions* 3/4, 79–93.

O'Shane, Patricia (1976) 'Is there any relevance in the women's movement for Aboriginal women?', *Refractory Girl* 28, 31–4.

Owens, Craig (1983) 'The discourse of others: Feminists and postmodernism' in *Postmodern Culture* (ed. H. Foster), 57–82, Pluto, London.

Pain, R. (1991) 'Space, sexual violence and social control: Integrating geographical and feminist analyses fear of crime', *Progress in Human Geography*, 15 (4), 415–31.

Pateman, Carole (1988). *The Sexual Contract*, Polity Press, Cambridge.

Pattison, W.D. (1964) 'The four traditions of geography', *Journal of Geography*, 63 (5), 211–16.

Pateman, Carole and Gross, Elizabeth (1986). *Feminist Challenges: Social and Political Theory*, Allen & Unwin, North Sydney.

Peake, Linda (1993) '"Race" and sexuality: challenging the patriarchal structuring of urban space', *Environment and Planning D* 11, 415–32.

Peel, Mark (1995). *Good Times, Hard Times. The Past and the Future in Elizabeth*, Melbourne University Press, Melbourne.

Permezel, Melissa (1992) 'Women, fear and public space: two Melbourne case studies', Unpublished BA (Hons) thesis, University of Melbourne.

Pettman, Jan (1992) *Living in the Margins. Racism, Sexism and Feminism in Australia*, Allen & Unwin, North Sydney, NSW.

Phillips, Anne and Taylor, Barbara (1980) 'Sex and skill: Notes towards a feminist economics', *Feminist Review*, 6, 79–88.

Pile, Steve and Rose, Gillian (1992) 'All or nothing? Politics and critiques in the modernism-postmodernism debate', *Environment and Planning D: Society and Space*, 10, 123–36.

Pinch, S. (1993) 'Social polarization: A comparison of evidence from Britain and the United States', *Environment and Planning A*, 25, 779–95.

Piore, M. and Sabel, C. (1984) *The Second Industrial Divide: Possibilities for Prosperity*, Basic Books, New York.

Pollert, Anna (1981) *Girls, Wives and Factory Lives*, Macmillan, London.

Poovey, Mary (1988) 'Feminism and deconstruction', *Feminist Studies*, 14 (1), 51–65.

Powell, Diana (1993) *Outwest: Perceptions of Sydney's Western Suburbs*, Allen and Unwin, Sydney.

Prakash, Gyan (1990) 'Writing post-Orientalist histories on the third world: Perspectives from Indian historiography', *Comparative Studies In Society and History*, 32

Pratt, Geraldine (1993) Reflections on poststructuralism and feminist empirics, theory and practice, *Antipode* 25 (1), 51–63.

Pratt, May Louise (1992) *Imperial Eyes. Travel Writing and Transculturation*, Routledge, London.

Queensland National Parks and Wildlife Service (1991) *Park Guide: Carnarvon National Park*, QNPWS Brochure.

Radcliffe, Sarah (1994) '(Representing) post-colonial women: Authority, difference and feminisms', *Area*, 26, (1), 25–32.

Raymond, Janice (1991) *A Passion for Friends: Toward a Philosophy of Female Friendship*, Beacon Press, Boston.

Ramazanoglu, C. (1986) 'Ethnocentrism and socialist feminist theory: A response to Barrett and McIntosh', *Feminist Review*, 22, 83–6.

Reekie, Gail (1993) *Temptations: Sex. Selling and the Department Store*, Allen & Unwin, St Leonards, NSW.

Rengert, A. and Monk, Janice (1980) *Overcoming Masculine Bias in Introductory Human Geography. A Module Approach*, Association of American Geographers, Washington, DC.

Reynolds, Henry (1972) *Aborigines and Settlers: The Australian Experience 1788–1939*, Cassell, Stanmore, NSW.

Reynolds, Henry (1990a) *Frontier, Aborigines, settlers and land*, Allen & Unwin, Sydney.

Reynolds, Henry (1990b) *With the White People*, Penguin, Ringwood, Vic.

Rich, Adrienne (1976) *Of Woman Born: Motherhood as Experience and Institution*, W.W. Norton, New York.

Rich, Adrienne (1980) 'Compulsory heterosexuality and lesbian existence', *Signs*, 5 (4), 631–60.

Riger, S. and Gordon, M. (1981) 'The fear of rape: A study of social control', *Journal of Social Issues*, 37 (4), 71–91.

Roberts, Marion (1991) *Living in a Man-Made World: Gender Assumptions in Modern House Design*, Routledge, London.

Rock, Cynthia, Torre, Susana and Wright, Gwendolyn (1980) 'The appropriation of the house: Changes in house design and concepts of domesticity' in *New Space for Women* (eds G. Wekerle, R. Peterson and D. Morley), 83–100, Westview Press, Boulder, Colorado.

Roder, Wolf (1977) 'An alternative interpretation of men and women in geography', *The Professional Geographer*, 29 (4), 397–400.

Rose, Gillian (1989) 'Locality studies and waged labour: an historical critique', *Transactions of the Institute of British Geographers*, 14, 317–28.

Rose, Gillian (1993) *Feminism and Geography: the Limits of Geographical Knowledge*, Polity Press, London.

Rose, Gillian (1995) 'The interstitial perspective: A review essay on Homi Bhabha's The Location of Culture', *Society and Space* 13, 365–73.

Rowbotham, Sheila (1972) *Women, Resistance and Revolution*, Penguin, Harmondsworth.

Rowland, Robyn (1992/3) *Living Laboratories: Women in Reproductive Technologies*, Indiana University Press, Bloomington.

Rowland, Robyn and Klein, Renate (1990) 'Radical feminism: Critique and construct' in *Feminist Knowledge: Critique and Construct* (ed. S. Gunew), 271–303, Routledge, London and New York.

Rowland, Robyn and Klein, Renate (1996) 'Radical feminism: history, politics, action' in *Radically Speaking: Feminism Reclaimed* (eds D. Bell and R. Klein), 9–36, Spinifex, North Melbourne.

Rowley, Charles D. (1970) *The Destruction of Aboriginal Society*, Penguin, Harmondsworth, UK.

Rowley, Charles D. (1971) *Outcasts in White Australia*, ANU Press, Canberra.

Rubin, Barbara (1979) '"Women in geography" revisited: Present status, new options', *The Professional Geographer*, 31, 125–34.

Said, Edward (1979) *Orientalism*, Vintage, New York.

Sassen, Saskia (1991) *The Global City*, Princeton University Press, Princeton.

Sassen, Saskia (1994) *Cities in a World Economy*, Pine Forge, London.

Sassen, Saskia (1998) *Globalization and its Discontents*, The New Press, New York.

Saunders, Rickie (1990) 'Integrating race and ethnicity into geographic gender studies', *Professional Geographer* 42 (2), 228–31.

Savage, Mike (1985) 'Capitalist and patriarchal relations at work: Preston cotton weaving, 1890–1940' in *Localities, Class and Gender* (The Lancaster Regionalism Group), 177–94, Pion, London.

Sayer, Andrew (1993) 'Postmodernist thought in geography: A realist view', *Antipode*, 25 (4), 320–44.

Sayers, J. Evans, M. and Redclift. N. (eds) (1987) *Engels Revisited: New Feminist Essays*, Tavistock, London.

Schaffer, Kay (1988) *Women and the Bush: Forces of Desire in the Australian Cultural Tradition*, Cambridge University Press, Cambridge.

Schepple, K.L. and Bart, P.B. (1983) 'Through women's eyes: defining danger in the wake of sexual assault', *Journal of Social Issues*, 39 (2), 63–80.

Scott, H. (1976) *Women and socialism: Experiences from Eastern Europe*, Allison and Busby, London.

Seager. L. and Olson, A. (1986) *Women in the World: an International Atlas*, Pan Books: London and Sydney.

Searle, Glen (1996) *Sydney as a Global City*, NSW Department of Urban Affairs and Planning, Sydney.

Seccombe, Wally (1973) 'The housewife and her labour under capitalism', *New Left Review*, 8, 3–24.

Seddon, George (1991) *Backyards and Beyond: the Australian Backyard*, State Library of Victoria, published in conjunction with the exhibition 'Backyards and Beyond', Melbourne.

Shields, Rob (ed.) (1992) *Lifestyle Shopping. The Subject of Consumption*, Routledge, London and New York.

Shohat, Ella (1996) 'Notes on the "post-colonial"', in *Postcolonial Theory. A Reader*, (ed. P. Mongia), Arnold, London.

Short, J.R. (1996) *The Urban Order: An Introduction to Cities, Power and Culture*, Blackwell, Cambridge, Mass.

Sibley, David (1995) *Geographies of Exclusion. Society and Difference in the West*, Routledge, London & New York.

Singer, Linda (1992) 'Feminism and postmodernism' in *Feminists Theorize the Political*, (eds Butler, J. and Scott, J.W.), Routledge, London and New York.

Slemon, Stephen (1996) 'Unsettling the empire: Resistance, theory for the second world' in *Postcolonial Theory. A Reader*, (ed. P. Mongia), 72–83, Arnold, London.

Smith, Neil (1994), 'Geography, empire and social theory', *Progress in Human Geography*, 18 (4), 491–500.

Smith, S.J. (1987) 'Fear of crime: Beyond a geography of deviance', *Progress in Human Geography*, 11 (1), 1–23.

Soja, Edward W. (1986) 'Taking Los Angeles apart: A postmodern geography', *Environment and Planning D: Society and Space*, 4, 255–72.

Soja, Edward W. (1989) *Postmodern Geographies: the Reassertion of Space in Critical Social Theory*, Verso, London.

Soja, Edward (1996) *Thirdspace: Journeys to Los Angeles and Other Real and Imagined Places*, Blackwell, Cambridge, Mass.

Spain, Daphne (1992), *Gendered Spaces*, University of North Carolina Press, Chapel Hill and London.

Spearritt, Peter (1995) 'Suburban cathedrals: The rise of the drive-in shopping centre' in *The Cream Brick Frontier: Histories of Australian Suburbia* (eds G. Davison, T. Dingle and S. O'Hanlon), 88–107, Monash Publications in History, 19, Clayton.

Spelman, Elizabeth (1988) *Inessential Woman. Problems of Essentialism is Feminist Thought*, Beacon Press, Boston.

Spender, Dale (1982) *Women of Ideas and What Men Have Done to Them*, Ark Paperbacks, London.

Spivak, Gayatri (1987) *In Other Worlds: Essays in Cultural Politics*, Methuen, London and New York.

Spivak, Gayatri (1990) *The Post-colonial Critic: Interviews, Strategies, Dialogues*, Routledge, London & New York.

Stanko, E, (1990) 'Everyday violence: How women and men experience sexual and physical danger', *Social Problems*, 32 (3), 238–50.

Stasiulis, Daiva and Yuval-Davis, Nira (1995) 'Introduction: beyond dichotomies—gender, race and class in settler societies' in *Unsettling Settler Societies* (eds D. Stasiulis and N. Yuval-Davis), 1–38. Sage, London.

Stilwell, Frank (1989) 'Structural change and spatial equity in Sydney', *Urban Policy and Research*, 7 (1), 3–14.

Sykes, Bobbie (1989) 'Blacks in the public sphere', *Hecate* 17, 51–3.

Sykes, Trevor (1993) 'Setting store by Westfield', *Australian Business Monthly*, May, 30–41.

Tanner, Leslie (ed.) (1970) *Voices from Women's Liberation*, New American Library, New York.

Taylor, Barbara (1983) *Eve and the New Jerusalem*, Virago, London.

Thomas, Pam and Skeat, H. (1990) 'Gender in third world development studies', *Australian Geographical Studies*, 28 (1), 5–15.

Tong, Rosemary (1989) *Feminist Thought: a Comprehensive Introduction*, London, Unwin Hyman.

Triffin, Helen (1990) 'Introduction' in *Past the Last Post: Theorizing Post-colonialism and Post-modernism*, viii–xvi, University of Calgary Press, Calgary.

Valentine, Gillian (1989) 'The geography of women's fear', *Area*, 21 (4), 385–90.

Valentine, Gillian (1990) 'Women's fear and the design of public space', *Built Environment* 16 (4), 288–303.

Valentine, Gillian (1992) 'Images of danger: Women's sources of information about the spatial distribution of male violence', *Area* 24, 22–9.

Valentine, Gillian (1993a) '(Hetero)sexing space: Lesbian perceptions and experiences of everyday spaces', *Environment and Planning D: Society and Space*, 11 (4), 395–413.

Valentine, Gillian (1993b) 'Desperately seeking Susan', *Area*, 25, 109–16.

Vogel, Lisa (1983) *Marxism and the Oppression of Women: Towards a Unitary Theory*, Pluto, London.

Walby, Sylvia (1985) 'Theories of women, work and employment' in *Localities, Class and Gender*, (The Lancaster Regionalism Group), 145–76, Pion, London.

Walby, Sylvia (1986) *Patriarchy at Work. Patriarchal and Capitalist Relations in Employment*, Polity Press, Oxford.

Walker, Megan (1999) 'Women—state of play', Melbourne *Age Good Weekend*, 6 March, 16–17.

Warde, A. (1991) 'Gentrification as consumption. Issues of class and gender', *Environment and Planning D: Society and Space*, 9 (2), 223–32.

Warr, Mark (1985) 'Fear of rape among urban women', *Social Problems*, 32 (3), 238–50.

Watson, Sophie (1988) *Accommodating Inequality: Gender and Housing*, Allen & Unwin, Sydney

Watson, Sophie (1990) *Playing the State: Australian Feminist Interventions*, Allen & Unwin, Sydney.

Weedon, Chris (1987) *Feminist Practice and Poststructuralist Theory*, Basil Blackwell, Oxford.

Westfield Holdings (1998) *Annual Report*, www.Connect com.au/ar/ar98.htm

Westfield Parramatta (1999) *Overview of Westfield Parramatta Trade Area*.

Westfield Parramatta (1999) *Market Areas and Key Socio-economic Characteristics*.

Westfield Parramatta (1999) *Sydney's Centre of Shopping*.

Westfield Parramatta (1999) *Directory and Services Guide*.

Westwood, Sally (1984) *All Day Every Day: Factory and Family in the Making of Women's Lives*, Pluto, London.

Wheelwright, Edward L. and Miskelly, J. (1967) *Anatomy of Australian Manufacturing Industry*, The Law Book Company, Sydney.

Wilson, Elizabeth (1989) *Hallucinations. Life in the Post-modern City*, Radius, London.

Wilson, Elizabeth (1991) *The Sphinx in the City. Urban Life, the Control of Disorder and Women*, Virago, London.

Wollstonecraft, Mary (1792/1929) *The Rights of Woman*, E.P. Dutton and Co., London and Toronto.

Women and Geography Study Group of the IBG (1984) *Geography and Gender*, Explorations in Feminism Collective in association with Hutchinson, London.

Wood, D. (1992) *The Power of Maps*, Guidlford Press, New York.

Worrall, Airlie (1987) 'All wool and a yard wide: The Victorian textile industry 1900–1930', Unpublished PhD thesis, University of Melbourne.

Wotherspoon, Gary (1991) *City of the Plain: History of a Gay Sub-culture. Sydney*, Hale and Iremonger, Sydney.

Wright, Gwendolyn (1980) *Moralism and the Model Home: Domestic Architecture and Cultural Conflict in Chicago 1973–1913*, University of Chicago Press, Chicago and London.

Yeatman, Anna (1995) 'Interlocking oppressions' in *Transitions. New Australian Feminisms*, 42–56, (eds B. Caine and R. Pringle), Allen & Unwin, St Leonards.

Yeoh, Brenda S.A. and Chang Tou Chuand (1995) 'The challenge of post-modern scholarship within geography', *Sojourn*, 10 (1), 116–30.

Zelinsky, Wilbur (1973a) 'The strange case of the missing female geographer', *The Professional Geographer*, 25 (2), 101–5.

Zelinsky, Wilbur (1973b) 'Women in geography: A brief factual account', *The Professional Geographer*

Zelinsky, Wilbur (1977) 'Vive la difference geographique? Further thoughts on geography as a male preserve', *The Professional Geographer*, 29 (4), 400–2.

Zelinsky, Wilbur, Monk, Janice and Hanson, Susan (1982) 'Women and geography. A review and prospectus', *Progress in Human Geography*, 6, 317–66.

Zukin, Sharon (1991) *Landscapes of Power. From Detroit to Disney World*, University of California Press, Berkeley.

■

Index